W9-AXW-093

Glossary of Weather and Climate

with Related Oceanic and Hydrologic Terms

Edited by
Ira W. Geer

American Meteorological Society
45 Beacon Street, Boston, Massachusetts 02108

The American Meteorological Society (AMS), founded in 1919, is a scientific and professional society. Interdisciplinary in its scope, the Society actively promotes the development and dissemination of information on the atmospheric and related oceanic and hydrologic sciences. AMS has more than 10,000 professional members from more than 100 countries and over 135 corporate and institutional members representing 40 countries.

This glossary was produced under the Project ATMOSPHERE initiative. Project ATMOSPHERE is the educational initiative of the American Meteorological Society to foster the teaching of atmospheric topics across the curriculum in grades K–12. It is a unique partnership between scientists and teachers with the ultimate goal of attracting young people to further studies in science, mathematics, and technology.

Project ATMOSPHERE has two major components. A national network of Atmospheric Education Resource Agents (AERAs) has been established and will have representation in every state. AERAs are primarily master precollege teachers who have been trained to engage in special leadership roles in their local and state educational systems and teacher associations. The other major component is the development and dissemination of scientifically accurate, up-to-date, and instructionally sound resource materials, such as this glossary, for teachers and students.

© 1996 American Meteorological Society
Permission to use *brief* excerpts from this book in scientific and educational works is hereby granted provided the source is acknowledged. All rights reserved. No part of this publication may be reproduced, stored in a retrieval system, or transmitted, in any form or by any means, electronic, mechanical, photocopying, recording, or otherwise, without the prior written permission of the publisher.

ISBN 1-878220-19-5
ISBN 1-878220-21-7 (paperbound)

Published by the American Meteorological Society
45 Beacon Street, Boston, MA 02108

Richard E. Hallgren, Executive Director
Keith L. Seitter, Associate Executive Director
Melissa I. Weston, Publications Coordinator

Printed in the United States of America
by Braun-Brumfield, Inc.,
A Sheridan Group Company

Glossary of Weather and Climate

with Related Oceanic and Hydrologic Terms

EDITOR

Ira W. Geer

PRINCIPAL CONTRIBUTORS

Kathryn M. Ginger

Joseph M. Moran

Edward J. Hopkins

Robert S. Weinbeck

David R. Smith

TABLE OF CONTENTS

Preface

This glossary is intended for educators, students, and the public. Its development was inspired by increasing contemporary interest in the atmosphere and global change and by the absence of a timely and reasonably comprehensive glossary for a general audience. The objective of the glossary is to provide a readily understandable, up-to-date reference for terms that are frequently used in discussions or descriptions of meteorological and climatological phenomena. In addition, the glossary includes definitions of related oceanographic and hydrologic terms.

This work started with a selection and updating of terms appearing in the seminal publication, *Glossary of Meteorology*, published by the American Meteorological Society (AMS) in 1959. A significant number of terms were taken directly or with modification from the *International Meteorological Vocabulary* (1992) of the World Meteorological Organization with the permission of Dr. G.O.P. Obasi, Secretary General. Additional terms for consideration were gleaned from examination of college-level meteorology textbooks and other publications. These were reviewed by several university faculty members, precollege teachers, and AMS staff in an effort to identify other relevant terms. To the extent possible, definitions were written or modified by meteorologists in an attempt to bring them to a more common descriptive level and style.

Definitions were then carefully reviewed and compared with definitions from a variety of sources listed elsewhere. Terms that relate to weather forecasts, watches, and warnings were provided or reviewed by U.S. National Weather Service personnel.

This glossary was made possible by the dedicated efforts of many people. Kathryn M. Ginger (AMS) was instrumental in compiling terms from the major sources, making preliminary modifications, and designing the format. Prof. Robert S. Weinbeck (State University of New York at Brockport) and Prof. David R. Smith (United States Naval Academy) reviewed initial versions and added terms. Prof. Joseph M. Moran (University of Wisconsin—Green Bay) reviewed every entry to update and check scientific accuracy. He also added numerous new terms, including many from oceanography and hydrology. Dr. Edward J. Hopkins (University of Wisconsin—Madison) conducted intensive reviews of later versions of the manuscript, revised definitions, and added a significant number of terms. National Weather Service input was arranged through Linda S. Kremkau. A reading of content was made by Prof. Smith and Thomas Wells (AMS), and Bernard A. Blair (AMS) made proof-reading corrections, as did Denise Taylor, AMS copy editor.

The order of entry in the glossary is alphabetical. Words with hyphens are treated as a continuation of the word.

Some of the terms are described using terms that have a particular meaning and appear elsewhere in the glossary. To aid in the use of this glossary, the following convention is used. Words or terms appearing in **bold type** are cross-referenced with defined terms found elsewhere in the glossary under that name. Some words are identified with *italics* to highlight a name or to indicate a specific meaning as described with the current entry, and no cross-reference search is necessary.

Several entries will refer to another definition that is a more preferred definition with notation such as ''same as . . .'' or ''see . . . .'' This scheme is meant to eliminate duplication. Some entries will contain the notation ''also known as . . . .''

Several entries include multiple definitions. For those within the same discipline, these entries are numbered, with the most common definition listed first. When the term is used in different contexts in several different disciplines, a notion will be made.

A listing of references and the International System (SI) of Units appear at the end of the glossary. Comments and suggestions are invited for consideration in future editions.

Ira W. Geer
AMS Education Program

A

ablation—The combined mass-depleting processes (such as **sublimation, melting**, **evaporation**) that remove snow or ice from the surface of a **glacier, snowfield**, etc.; also the amount of snow and ice lost by this process. Opposite of **accumulation**.

ablation zone—That portion of a **glacier** below the **firn line**, where **ablation** exceeds **accumulation**. Opposite of **accumulation zone**.

absolute altitude—The true or exact vertical distance above the earth's surface (**AGL**). Often obtained from the **radar altitude**. Contrast with **true altitude**.

absolute humidity—In **humid air**, the ratio of the mass of water vapor present to the volume occupied by the mixture of water vapor and dry air; that is, the *density* of the water vapor component. Normally expressed as grams of water vapor in a cubic meter of air.

absolute instability—The state of a column or layer of air in the **atmosphere** when it has a **superadiabatic lapse rate** of temperature (i.e., greater than the **dry-adiabatic lapse rate**). An **air parcel** displaced vertically would be accelerated in the direction of the displacement because of buoyancy considerations. See **parcel method**.

absolute salinity—A measure of the concentration of dissolved salts in **seawater** (**salinity**) based upon electrical conductivity. Values of absolute salinity are defined as 1.80655 times the **chlorinity**. Compare with **practical salinity**.

absolute stability—The state of a column or layer of air in the **atmosphere** when its **lapse rate** of temperature is less than the **saturation-adiabatic lapse rate**. An **air parcel** displaced vertically by an **adiabatic process** would tend to return to its level of origin because of buoyancy considerations. See **parcel method**. Contrast with **conditional stability**.

absolute temperature scale—Same as **Kelvin temperature scale**.

absolute vorticity—A measure of the spin of a **fluid** (**vorticity**) relative to an absolute reference system. Usually, the vertical component of vorticity is implied. The absolute vorticity is the sum of the vertical component of the **vorticity** with respect to the earth (the **relative vorticity**) and the vorticity of the earth (the **Coriolis parameter**).

absolute zero—The zero point of the **Kelvin temperature scale**; of fundamental significance in **thermodynamics**. It may be interpreted as the temperature at which the volume of a **perfect gas** vanishes. The value of absolute zero (0 K) on the **Celsius temperature scale** is estimated to be –273.16°C ± 0.01 (–459.67°F).

absorption—(1) The penetration and incorporation of one object into another. Contrast with **adsorption**. (2) The process through which incident radiant energy is retained by a substance. The absorbed radiation is then converted to another form of energy (e.g., heat).

absorptivity—A measure of the amount of **radiant energy** absorbed by a given substance; defined as the ratio of the amount of radiant energy absorbed to the total amount incident upon that substance. The absorption may occur either at the surface of (as with an opaque body) or in transit through the substance. A function of the material, its surface, and the wavelength of the incident radiation.

Ac—Abbreviation for the cloud type ***altocumulus.***

acceleration—The rate of change with time of the speed and/or direction of a moving object; strictly, the rate of change with time of the velocity vector of a particle. A decrease in speed is sometimes called *deceleration.* Units are length per time squared.

accessory cloud—A **cloud** form that is dependent, for its formation and continuation, upon the existence of one of the major **cloud genera**. It is often an appendage of the parent cloud (as **mamma**, **arcus**), but also may be an immediately adjacent cloudy mass (as **pileus**).

acclimatization—The gradual process through which a living organism becomes adapted to a change in environment. See **climatization.**

accretion—In cloud physics, the growth of a precipitation particle by the collision of a frozen particle (**ice crystal** or **snowflake**) with a supercooled liquid droplet that freezes upon contact. This is a form of **agglomeration** and is analogous to **coalescence**, in which both particles are liquid. See **coagulation**.

accumulation—In glaciology, the quantity of snow or other solid form of water added to a **glacier** or **snowfield** by **deposition** or **precipitation**. Opposite of **ablation**. See **snow accumulation**.

accumulation zone—The portion of a **glacier** above the **firn line**, where the accumulation exceeds **ablation**. Opposite of **ablation zone**.

acid deposition—The combination of **acid rain** (or snow) plus **dry deposition**.

acid rain—Rain having a **pH** lower than 5.6, representing the pH of natural rainwater; the increased acidity is usually due to the presence of sulfuric acid and/or nitric acid, often attributed to anthropogenic sources.

acoustic sounding—Study of the properties and structure of the atmosphere with the aid of observation of the passage of bursts of high-intensity **sound waves** sent upward and returned to a receiver.

acoustic wave (or *sound wave*)—A three-dimensional disturbance passing through an elastic or **fluid** medium consisting of a compression and a decompression of the medium. The wave moves through the medium at the "**speed of sound**" of that material. Humans can detect those waves that are within the **audio frequency** range.

acre-foot—A hydrologic quantity defined as the quantity of water required to cover 1 acre to a depth of 1 foot. One acre-foot is equivalent to 43 560 cubic feet, 325 851 gallons, or 1233 cubic meters.

active front—A **front**, or portion thereof, that produces appreciable cloudiness and, usually, precipitation.

active layer—The part of the soil that usually freezes in winter and thaws in summer. Its bottom surface is the **frost table**, beneath which may lie **permafrost**. The depth of the active layer varies anywhere from a few inches to several feet.

actual evapotranspiration (also known as *effective evapotranspiration*)—The quantity of water evaporated from soil and plants when the ground is at its natural moisture content. This is the amount measured by a **lysimeter**. Contrast with **potential evapotranspiration**.

adiabat—In most contexts, same as **dry adiabat**.

adiabatic atmosphere—A **model atmosphere** characterized by a **dry-adiabatic lapse rate** throughout its vertical extent.

adiabatic chart—Often used synonymously with **thermodynamic diagram**.

adiabatic cooling—An expansional cooling of an **air parcel** in an **adiabatic process**.

adiabatic equilibrium—A vertical distribution of temperature and pressure in an atmosphere in **hydrostatic equilibrium** such that an **air parcel** displaced following an **adiabatic process** will continue to possess the same temperature and pressure as its surroundings, so that no restoring force acts on a parcel displaced vertically. Such a condition is reached in a layer experiencing strong vertical **mixing**.

adiabatic lapse rate—A theoretical **process lapse rate**. In many contexts, same as **dry-adiabatic lapse rate**.

adiabatic process—A thermodynamic **change of state** of a system in which no heat or mass is transferred across the boundaries of the system. In an adiabatic process, *compression* always results in warming, and *expansion* in cooling. In meteorology the adiabatic process is often also taken to be a **reversible process**. Vertical motions of **air parcels** in the atmosphere are approximately adiabatic.

adiabatic warming—A compressional warming of an **air parcel** in an **adiabatic process**.

adsorption—The adhesion of a thin film of **liquid** or **gas** to a **solid** substance. The solid does not chemically combine with the adsorbed substance. Contrast with **absorption** (1).

Advanced Weather Interactive Processing System (**AWIPS**)—A computerized workstation that enables National Weather Service meteorologists to integrate a variety of weather data with displays of graphics and satellite and radar imagery.

advection—Horizontal **transport** of an atmospheric property (e.g., **temperature**, **moisture,** or **vorticity**) solely by the **wind**; also, the rate of change of the value of the advected property at a given point due solely to wind transport. See **cold air advection** or **warm air advection**.

advection fog—A type of **fog** caused by the movement (**advection**) of mild, **humid air** over a cold surface, and the consequent cooling of that air to below its initial **dew-point**. A very common advection fog is that caused by **humid air** in transport over a cold body of water (**sea fog**).

advection frost—The occurrence of **frost** as a result of the horizontal **transport** (**advection**) of a cold air mass with air temperatures below 0°C (32°F) by the **wind**; may also refer to advection of moisture at subfreezing temperatures. This type of frost is responsible for causing damage to agricultural areas of south Florida and the Rio Grande Valley of Texas during cold **polar outbreaks**. Contrast with **radiation frost**.

advective thunderstorm—A **thunderstorm** resulting from **instability** produced by the differential **transport** (**advection**) of relatively colder air at high levels or relatively warmer air at low levels, or by a combination of both conditions.

advisory forecast—Any weather forecast issued for special purposes. See **weather advisory**. The National Weather Service issues advisories to the public to highlight potential or current weather conditions that require caution but are deemed to be less serious than those that would warrant issuance of a **watch** or **warning**. The events described in an advisory can cause significant inconvenience and if caution were not exercised, could lead to situations that may threaten life and/or property.

aeolian—Pertaining to the action or the effect of the **wind**.

aeration—In general, any process whereby a substance becomes permeated with air or other **gas**. Usually, dissolved oxygen level of the water is increased by this process.

aerodynamic—Pertaining to forces acting upon any **solid** or **liquid** body moving relative to a **gas** (especially air); also *aerodynamics*, the scientific study of compressible **fluid** flow.

aeronomy—The study of the physics of the **upper atmosphere**. It is concerned with upper-atmospheric composition (i.e., nature of constituents, density, temperature, etc.) and chemical reactions.

aerosol—A **colloidal system** in which the *dispersed phase* is composed of either **solid** or **liquid** particles, and in which the *dispersion medium* is some **gas**, usually air. Examples include **smoke**, **dust,** and **fog**.

aestival—Pertaining to **summer**. The corresponding adjectives for **autumn**, **winter**, and **spring** are **autumnal, hibernal,** and **vernal**.

afterglow—A broad, high arch of **radiance** seen occasionally in the western sky above the highest clouds in deepening evening **twilight** when the sun is 3°–4° below the horizon. It is caused by the **scattering** effect exerted upon the components of white light by very fine particles suspended in the high atmosphere. When used in this rather broad sense, the term embraces all of the complex luminosity phenomena observed in the western twilight sky. See also **alpenglow**.

age—In general, a time interval, with specific uses to include: (1) In geology, a **geologic age**—a unit of **geologic time** smaller than a **geologic epoch**. (2) In astronomy, the elapsed time from the start of some phenomena, for example, the *"age of the* *moon.*'' (3) In oceanography, the time interval between an astronomical event and the corresponding tidal phenomenon. For example, the lag in days between the occurrence of **syzygy** and the highest **spring tide** is the ''*age of the tide*''; the lag between **perigee** and the highest **perigee tide** is the ''*age of parallax inequality.*''

ageostrophic—A condition of wind not satisfying the **geostrophic** approximation. The departure of the actual wind from the **geostrophic wind**.

agglomeration—In cloud physics, the process through which precipitation particles grow by **collision** with and assimilation of cloud particles or other precipitation particles.

AGL—An abbreviation for *a*bove *g*round *l*evel. Contrast with **MSL** or above **mean sea level**.

agricultural climatology—In general, the scientific study of the effect of climate on crops; **applied climatology**. It includes especially the length of the **growing season**, the relation of growth rate and crop yields to the various climatic factors and hence the optimum and limiting climates for any given crop, the value of **irrigation**, and the effect of climatic and weather conditions on the development and spread of crop diseases.

agricultural meteorology—In general, **meteorology** and **micrometeorology** as applied to the specific problems of agriculture.

Agulhas Current—A warm, generally southwestward-flowing ocean current of the southwestern Indian Ocean some 200 kilometers off the coast of South Africa; one of the swiftest of ocean currents; contains the **Mozambique Current**. It is fed by the **South Equatorial Current** for part of the year and merges to contribute to the **South Indian Current**.

air—A mixture of **gases** consisting of mostly nitrogen (N_2) and oxygen (O_2), and constituting the earth's **atmosphere**, together with suspended solid and liquid **aerosols**. See **dry air** and **humid air**.

air discharge—A **lightning discharge** between a cloud and a cloud-free (or clear air) region of the atmosphere.

air drainage—General term for **gravity**-induced, downslope flow of relatively cold air; a form of **katabatic wind**. Winds thus produced are called **gravity winds**.

air mass—A widespread body of air, the thermal, moisture, and stability properties of which can be identified as (*a*) having been established while that air was situated over a particular region of the earth's surface (climatology), and (*b*) undergoing specific modifications while in transit away from the **source region**. An air mass is often defined as a widespread body of air extending over an area of several million square kilometers that is approximately homogeneous in its horizontal extent, particularly with reference to temperature and water vapor concentration; in addition, the vertical temperature and humidity variations, together with the stability, are approximately the same over its horizontal extent. Depending upon its type, the depth of the air mass may range from one to several kilometers.

air parcel—An imaginary volume of air to which may be assigned any or all of the basic **dynamic** and **thermodynamic** properties of atmospheric air; often used as a convenient tracer of air movement or atmospheric processes.

air pollution—Same as **atmospheric pollution**.

air pressure—Same as **atmospheric pressure**.

air quality—A measure of the cleanliness of air described in terms of levels of **contaminants** in air especially with regard to their potential effects on human health.

air quality standards—A set of objectives established internationally by the World Health Organization and nationally by the U.S. Environmental Protection Agency for limiting the levels of certain air pollutants. See **National Ambient Air Quality Standards** (**NAAQS**).

aircraft ceiling—(1) In United States **aviation weather observation** practice, the height of the lowest layer of clouds as determined by a pilot while in flight within one and one-half nautical miles of any runway of the airport. (2) The maximum altitude at which any given aircraft can be operated safely.

aircraft icing—Formation of a deposit of **ice**, **rime,** or **hoarfrost** on surfaces of aircraft; this situation can be hazardous, especially when in flight. **Pilot reports** and **aviation weather forecasts** provide categorical descriptions of the intensity of aircraft frame icing based upon accumulation rates of ice on the aircraft.

aircraft report (**AIREP**)—Same as **pilot report**.

aircraft sounding—The vertical distribution of air pressure, temperature, and humidity data observed by instruments onboard an aircraft during climb or descent of the aircraft. Compare with **aircraft report**.

aircraft turbulence—Irregular and chaotic atmospheric flow (**turbulence**) that may be hazardous to aircraft operation. **Pilot reports** and **aviation weather forecasts** provide categorical descriptions of turbulence intensity based upon aircraft reaction or the reaction of occupants in the aircraft.

aircraft weather reconnaissance—The making of detailed weather observations or investigations from aircraft in flight; often used to describe a mission to reconnoiter a **tropical cyclone** (**hurricane/tropical storm/tropical depression**).

AIREP—Abbreviation for **aircraft report**. See **pilot report**.

airglow—The quasi-steady, faint, radiant emission from the **upper atmosphere** over middle and low latitudes; as distinguished from the sporadic emission of **aurorae** that occur over high latitudes.

airmass classification—Identification of the various **air masses** on a **synoptic chart** by the determination of their physical characteristics and development using a standard scheme. The classification scheme most often uses the thermal properties of the **airmass source region**, with additional classification based upon moisture characteristics. It constitutes an important part of weather map analysis. See also **arctic air**, **polar air**, and **tropical air**.

airmass climatology—A statistical based analysis of climate of a locale based upon the frequency of occurrence and characteristics of the various types of **air masses** affecting the locale. A type of **synoptic climatology**.

airmass shower—Intermittent **precipiation** produced by local **convection** due to surface heating and found in a moist, unstable **air mass**. **Airmass thunderstorms** are the extreme.

airmass source region—An extensive and homogenous area of the earth's surface over which large bodies of air frequently remain for a sufficient time to acquire characteristic temperature and humidity properties imparted by that surface. These regions may be large oceanic or land areas. Air so modified becomes identifiable as a distinct **air mass**.

airmass thunderstorm—A type of **thunderstorm** that forms in a warm, moist, unstable **air mass** and is not associated with frontal or associated synoptic-scale lifting mechanisms. Typically, this type of thunderstorm forms in a tropical air mass during the late afternoon, is not organized, and does not reach the criteria of a **severe thunderstorm**.

airport elevation (also known as *field elevation*)—The officially designated **elevation** of an airport above mean sea level. It is the elevation of the highest point on any of the runways of the airport. This elevation usually is the same as **station elevation**, if **air pressure** were measured at the airport.

air–sea interaction—The interchange of **energy** (e.g., heat, kinetic energy) and **mass** (e.g., moisture, particles) that takes place across the active surface **interface** between the top layer of the ocean and the layer of air in contact with it and vice versa.

airspeed—The speed of an exposed (usually airborne) object relative to the atmosphere. In a **calm** atmosphere, airspeed is equal to **ground speed**.

airstream—Air in motion, usually refers to an **air mass** moving away from its **source region**.

Aitken nucleus—A microscopic atmospheric particle with a diameter less than 0.4 micrometer. These particles can serve as **condensation nuclei** for droplet formation under conditions of rapid adiabatic expansion, such as in a cloud chamber. Named for John Aitken (1839–1911), a Scottish physicist who studied condensation processes in the late nineteenth century.

Alaska Current—A relatively warm ocean current in the North Pacific Ocean; the northward-flowing segment of the **Aleutian Current**. It circulates **cyclonically** around the Gulf of Alaska; part of the water passes between the Aleutian Islands into the Bering Sea from which it emerges as the **Oyashio**, and part rejoins the Aleutian Current.

albedo—The ratio of the amount of **electromagnetic radiation** reflected by a body to the amount incident upon it; commonly expressed as a percentage. Usually, albedo refers to radiation in the **visible** range or to the full spectrum of **solar radiation**; **reflectivity** usually refers to radiation in any narrow wavelength band. See **planetary albedo**.

Alberta low—A **cyclone** or an atmospheric low pressure cell centered on the eastern slope of the Canadian Rockies in the province of Alberta, Canada. Formerly, such lows were thought to have originated (more or less independently) over this location. Depressions moving inland from the Pacific are now recognized as the actual parent systems. Alberta lows appear as these systems enhance, or are enhanced by, the **dynamic trough** that is a typical, almost semipermanent, feature of this region. See **Colorado low**.

alcohol thermometer—A **liquid-in-glass thermometer** containing an alcohol such as ethanol. Because of its low freezing point, alcohol is used in **minimum thermometers**, especially those used in northern latitudes where temperatures routinely reach –40°C (–40°F).

Aleutian Current (also known as *Subarctic Current*)—A cold eastward-flowing ocean current that lies north of the **North Pacific Current** near the Aleutian Islands; it is the northern branch of the **Kuroshio Extension**, which moves northeast then east between 40° and 50° N. As it approaches the coast of North America, it divides to form the northward-flowing **Alaska Current** and the southward-flowing **California Current**.

Aleutian low—The quasi-permanent atmospheric low pressure center located near the Aleutian Islands (near 50° N) on charts of mean sea level pressure in winter. This low represents one of the main **centers of action** in the atmospheric circulation of the Northern Hemisphere.

alkalinity—In **seawater**, the excess of hydroxyl ions over hydrogen ions, generally expressed as milliequivalents per liter; having a **pH** in excess of 7.

almanac—A calendar to which astronomical and other data (often weather predictions) are commonly added.

aloft—A broad description of a location in the atmosphere at some height above the earth's surface; typically used in conjunction with an **upper-air observation** to distinguish from **surface weather observation**—such as **winds aloft**.

alpenglow—A reappearance of pink or yellow sunset colors on a mountain summit opposite the sun; sometimes observed after the original colors have faded; also, a similar phenomenon preceding the regular coloration at sunrise. See **afterglow**.

alpha particle—A positively charged nuclear particle consisting of two protons and two neutrons—that is, a helium nucleus. These particles are involved with atmospheric **ionization**. Compare with a **beta particle** and **gamma rays**.

alpine tundra—A form of **tundra** in which the absence of trees is due to high altitude instead of high latitude. It lies roughly between the summer isotherm of 10°C (50°F) and the **snow line** (1).

Alter shield—A type of **rain gauge shield**; consists of freely hanging, spaced slats, arranged circularly around a rain gauge. The advantage of this shield is that the slats do not easily accumulate snow, thus permitting its use on unattended gauges.

altimeter—An instrument that determines the **altitude** (1) of an object with respect to a fixed level; examples include **pressure altimeters** and **radio altimeters**.

altimeter corrections—Systematic adjustments that must be made to the **indicated al-**

titude readings of a **pressure altimeter** to obtain **true altitudes**. These corrections are necessary for two types of errors: (*a*) horizontal pressure gradient error resulting from horizontal changes in an existing **altimeter setting** at the surface; and (*b*) air temperature error resulting from the difference between the actual atmospheric temperature of the air column extending from beneath the airplane to sea level and that of the assumed **standard atmosphere**.

altimeter setting—The value of **atmospheric pressure** to which the scale of a **pressure altimeter** is set to indicate the **true altitude** of an aircraft at the **airport elevation**. In United States practice, the setting represents the pressure required to make the altimeter indicate zero altitude at an elevation of 10 feet (3 meters) above mean sea level. Thus, at the height of 10 feet above airport elevation (approximate cockpit height), the altimeter should indicate the airport elevation.

altitude—(1) The vertical distance above a reference level, typically **mean sea level**, or sometimes above ground level. The point or plane does not need to be connected to local surface, as in **elevation**. See **absolute altitude**, **corrected altitude**, **density altitude**, **indicated altitude**, **pressure altitude**, and **true altitude**. (2) Sometimes refers to **altitude angle**.

altitude angle—The angular distance of a celestial body above or below the local horizon, measured (in degrees) along a **great circle** on the **celestial sphere** containing both the body and the local **zenith**; called **elevation angle** for angles above the horizon. Altitude angle is 90° minus the **zenith angle**.

altocumulus (Ac)—A principal **middle cloud** type (**cloud genus**), white and/or gray in color, that occurs as a layer or patch with a waved aspect, the elements of which appear as laminae, rounded masses, rolls, etc. These elements usually are sharply outlined, but they may become partly fibrous or diffuse; they may or may not be merged; they generally have shadowed parts.

altocumulus castellanus (formerly known as *castellatus*)—A **cloud species** of the genus **altocumulus**, representing a **middle cloud** that appears to be vertically developed, with castle-like turreted protuberances. The cloud indicates **instability** and **turbulence** aloft between 6500 and 20 000 feet (2 and 6 kilometers).

altostratus (As)—A principal **middle cloud** type (**cloud genus**) in the form of a gray or bluish (never white) sheet or layer of striated, fibrous, or uniform appearance. Altostratus very often totally covers the sky, and may, in fact, cover an area of several thousand square miles. The layer has parts thin enough to reveal the position of the sun; and if gaps and rifts appear, they are irregularly shaped and spaced.

ambient air—The surrounding undisturbed outside air.

ambient air quality standards—The lowest allowable concentrations of those air pollutants regulated by federal law; applies to outside air only.

ambient temperature—The temperature of the surrounding air, typically measured with a **thermometer**; often equivalent to the **dry-bulb temperature**.

amorphous frost—**Hoarfrost** that possesses no apparent simple crystalline structure.

amount of precipitation—Depth to which **precipitation** would cover, in a **liquid**

form, on a horizontal projection of the earth's surface, in the absence of **infiltration**, **runoff**, or **evaporation** and if all solid precipitation were melted. See **snowfall** (2).

amplitude—The magnitude of the maximum displacement of a periodic wave from its mean value; or equivalently, one-half of the difference between the maximum and minimum values in a cycle.

anabatic wind—An upslope wind; usually applied only when the wind is blowing up a hill or mountain as the result of local surface heating, and apart from the effects of the larger-scale circulation. Opposite of **katabatic wind**. The most common anabatic wind type is the **valley wind**.

analog method—A forecasting method based on the assumption that a current **synoptic** weather situation will develop and migrate in the same manner as a similar case study synoptic pattern in the past.

analysis—In **synoptic meteorology**, a detailed diagnostic study of the state of the atmosphere based on actual weather observations, usually including a separation of the entity into its component patterns through the drawing of families of **isopleths** for various elements such as air pressure and temperature, or through **airmass classification**. Two types of analysis techniques are used: **objective** and **subjective analysis**.

anchor ice—Submerged ice attached to the beds of streams, lakes, and shallow seas or to other underwater objects, regardless of its nature of formation.

anemometer—The general name for instruments designed to measure the **wind speed**. Types of anemometers include **hot wire**, **rotation**, and **pressure-type anemometers**.

anemometer level—The recommended height at which an **anemometer** should be located above ground level for proper exposure; usually 10 meters (30 feet). The top of the **surface boundary level** is loosely defined at this level.

aneroid barometer—An instrument for measuring **atmospheric pressure**. It is constructed on the following principles: an *aneroid capsule* (a thin corrugated hollow disk) is partially evacuated of **gas** and restrained from collapsing by an external or internal spring. The deflection of the spring will be nearly proportional to the difference between the internal and external (atmospheric) pressures.

angle of incidence—The angle at which a ray of energy (or object) impinges upon a surface, measured between the direction of propagation of the energy (or object) and a perpendicular to the surface at the point of impingement, or incidence.

angle of reflection—The angle at which a reflected ray of energy leaves a reflecting surface, measured between the direction of the outgoing ray and a perpendicular to the surface at the point of reflection.

angle of refraction—The angle at which a refracted ray of energy leaves the interface at which the **refraction** occurred, measured between the direction of the refracted ray and a perpendicular to the interface at the point of refraction.

Ångstrom (Å)—A unit of length traditionally used in the measurement of the **wave-**

length of extremely short **electromagnetic radiation**, such as **X-rays**, and in the measurement of molecular and atomic diameters. Named for A.J. Ångstrom (1814–1874), a Swedish physicist. The current trend is to use **nanometers**. One Ångstrom equals 10^{-10} meters or 0.1 nanometers.

angular momentum—In terms of a unit mass, the product of the instantaneous linear **velocity** of a body rotating about an axis and its perpendicular distance from the axis. The absolute angular momentum of an **air parcel**, per unit mass, is the sum of its angular momentum relative to the earth and the angular momentum due to the earth's rotation. Units are $mass \times length^2/per\ unit\ time$. See (**linear**) **momentum**.

angular velocity—A measure of the rotation rate of a particle moving about a rotation axis. A **vector quantity**, whose magnitude is equal to the time rate of angular displacement (revolutions per unit time or degrees of arc per unit time) and whose direction is parallel to the axis of rotation, oriented in accordance with the *''right-hand rule''* (palm of right hand curls around axis in the same sense as rotation, and thumb points in direction of the resultant vector). For example, the angular velocity vector of the earth parallels its spin axis, pointing toward the North Pole, and with a magnitude of 360° of arc per **sidereal** day.

anniversary winds—General term for **local winds** or larger-scale wind systems (such as the monsoon) that recur annually.

annual—Yearly, especially pertaining to periodic variations that are completed within a **year**, or 365 days. Compare with **diurnal**.

annual flood—The highest flow (**peak discharge**) at a point on a stream during any particular calendar year or **water year**. Flow may be expressed as maximum instantaneous **stage** or **discharge**, or the highest 24-hour average. The annual flood need not exceed the established **flood stage**.

anomalous propagation—(1) The transmission of various forms of energy through the atmosphere when the energy is **refracted** to move along a path other than the line of sight path by density discontinuities at one or more levels in the atmosphere. It results from the unusual vertical **profiles** of temperature and humidity. Also known as **false echoes** or *angel echoes*. (2) Propagation of **sound waves** to distances greater than can be accounted for by straight-line propagation.

anomaly—(1) The deviation of (usually) temperature or precipitation in a given region over a specified period from the long-term average value for the same region. (2) In oceanography, the difference between conditions actually observed at a station and those that would have existed had the water all been of a given arbitrary temperature and **salinity**.

antarctic air—A type of cold, dry **air mass** developed over the Antarctic continent. Antarctic air appears to be colder at the surface in all seasons, and at all levels in austral autumn and winter, than **arctic air**.

antarctic anticyclone—The permanent atmospheric high pressure system assumed to overlie the continent of Antarctica; characterized by a circulation regime with inflow in high **troposphere**, **subsidence**, and low-level **anticyclonic** outflow near the surface; analogous to the **Greenland anticyclone**.

Antarctic Bottom Water—A cold saline **water mass** that sinks to a depth exceeding 12 000 feet (3600 meters) in the South Atlantic, near Antarctica; this water then flows northward to approximately 45°N. Typically, temperatures are about –0.4°C and **salinity** is 34.66 psu.

Antarctic Circle—The line of latitude 66° 33'S (often taken as 66.5°S). Along this line the sun does not set on the day of the austral **summer solstice**, about 22 December, and does not rise on the day of the **winter solstice**, about 21 June. From this line, the number of 24-hour periods of continuous daylight or of continuous night increases southward to about six months each at the geographic South Pole.

Antarctic Circumpolar Current (also known as the *west wind drift*)—The ocean current with the largest volume **transport** (approximately 110 **Sverdrups** or 110 million cubic meters per second), and the swiftest current; it flows from west to east through all the ocean surrounding the Antarctic continent. It is locally deflected from its course by the distribution of land and sea, the submarine topography, and the currents in the adjacent oceans. On its northern edge it is continuous with the **South Atlantic Current**, the **South Pacific Current**, and the eastward-flowing extension of the **Agulhas Current** in the Indian Ocean.

Antarctic Convergence—The zone in the Southern Oceans near 50°S marking the confluence of subantarctic and subtropical surface **water masses**. This zone is marked by a strong horizontal **temperature gradient**. **Antarctic Intermediate Water** is formed.

antarctic front—The semipermanent, semicontinuous **front** (at about 60° – 65°S) between the cold **antarctic air** of the antarctic continent and the relatively warm **polar air** of the southern oceans; generally comparable to the **arctic front** of the Northern Hemisphere.

antarctic high—Same as **antarctic anticyclone**.

Antarctic Intermediate Water—A cold less saline **water mass** formed in the surface layers off the antarctic continent. It sinks in the **Antarctic Convergence** and flows north at a depth of about 900 meters beneath the South Atlantic **gyre**.

antarctic ozone hole—A large area of intense stratospheric **ozone** depletion over the Antarctic continent that has been found to occur annually between late August and early October, and generally ends in mid-November. This severe ozone thinning is attributed to the action of chlorine liberated from a group of chemicals known as **chlorofluorocarbons (CFCs)**.

antarctic stratospheric vortex—Persistent westerly atmospheric circulation in the **stratosphere** during winter in the Southern Hemisphere, which is most intense between 60° and 70°S and increases with altitude up to the **stratopause**.

Antarctic Zone—Geographically, the region between the **Antarctic Circle** (66° 33'S) and the South Pole. Climatically, the limit of the zone may be set at about 60° S, poleward of which the prevailing westerly winds give place to easterly or variable winds. Over most of this region, the average temperature does not rise above 0°C (32°F) even in summer.

anticorona—Same as **glory**.

anticyclogenesis—Any development or strengthening of an **anticyclonic** circulation in the atmosphere. Opposite of **anticyclolysis**. Anticyclogenesis applies to the development of anticyclonic circulation (or the initial appearance of a **high**) where previously it was nonexistent, as well as to intensification of existing anticyclonic flow.

anticyclolysis—Any weakening of an **anticyclonic** circulation in the atmosphere. Opposite of **anticyclogenesis**. Anticyclolysis (which refers to the circulation) should be distinguished from **deepening** (which refers to a decrease in the atmospheric pressure), although the two processes usually occur together.

anticyclone—A dome of air that exerts relatively high atmospheric pressure compared with the surrounding air at a given level. It has a closed **anticyclonic** wind circulation. Essentially the same as a **high**. In the Northern Hemisphere, surface winds in an anticyclone blow clockwise and outward around the center of highest pressure; in the Southern Hemisphere a counterclockwise outflow occurs.

anticyclonic—Having a sense of rotation about the local vertical that is clockwise in the Northern Hemisphere and counterclockwise in the Southern Hemisphere. Opposite of **cyclonic**.

anticyclonic shear—Horizontal **wind shear** of such a nature that it contributes to the anticyclonic **vorticity** of the flow; that is, it tends to produce anticyclonic rotation of the individual air particles along the line of flow. In the Northern Hemisphere, anticyclonic shear is present if the **wind speed** decreases from left to right across the direction of flow; the opposite is true in the Southern Hemisphere. Contrast with **cyclonic shear**.

Antilles Current—An ocean current in the North Atlantic, the northern branch of the **North Equatorial Current** flowing along the northern side of the Great Antilles carrying water that is identical with that of the **Sargasso Sea**. The Antilles Current eventually joins the **Florida Current** (after the latter emerges from the Straits of Florida) to form the **Gulf Stream**.

antinode—The point in a **standing wave** where the vertical displacement of the water surface is the greatest.

antisolar point—The location on the **celestial sphere** (above or below the local **horizon**) that is directly opposite the sun from the observer (i.e., on the line from the sun through the observer). The observer's shadow is cast on any object in the direction of the person's antisolar point.

antitrades—A deep layer of **westerly** winds in the **troposphere** above the surface easterly **trade winds** of the Tropics and subtropics, on the equatorward side of the midlatitude westerlies; the term antitrades is used in context with the **Walker Circulation** cell.

antitwilight arch—The pink or purplish band of about three degrees vertical angular width that lies just above the **antisolar point** at **twilight**; it rises with the antisolar point at sunset and sets with the antisolar point at sunrise. This band is produced by **scattering** of sunlight from atmospheric materials in the high atmosphere.

anvil cloud(also known as ***incus***)—The popular name applied to the upper flattened portion of a **cumulonimbus** cloud formed as the **updraft** encounters a layer of stable air and begins to spread laterally in all directions in the vicinity of the **tropopause**. The anvil is a **cirriform** cloud, formed when the upper portion of the cumulonimbus becomes *glaciated*. See **glaciation** (3).

aphelion—The point on the annual orbit of the earth (or any other body in orbit about the sun) that is farthest from the sun; at present, the earth reaches this point (152 million kilometers from the sun) on about 5 July. Opposite of **perihelion**.

apogean tide—A tide of decreased **tidal range** occurring when the moon is near **apogee**. Opposite of **perigean tide**.

apogee—That point on the orbit of the moon (or any other earth satellite) that is farthest from the earth. Opposite of **perigee**.

apparent solar day—The interval of time between two successive transits of the true sun across a local **meridian**. This interval is about four minutes longer than the **sidereal day**, largely because of the sun's apparent annual motion eastward along the ecliptic (actually, the earth's "westward" motion along its orbit), which delays the sun's return to **meridional transit**. This interval is inconveniently nonuniform due to systematic variations in the earth's orbital speed around the sun, and due to the sun's changing **declination.** See **apparent solar time**. The concept of the **mean solar day** has been invented to circumvent these practical difficulties.

apparent solar time—The time reckoned upon the diurnal motion of the true sun across the **celestial sphere**. Daily motion of the actual sun undergoes variations throughout the year because of the earth's orbital **eccentricity** and the **obliquity of the ecliptic**. Contrast with **mean solar time**. Differences between apparent and mean solar time, as specified by the **equation of (ephemeris) time** may reach as much as 15 minutes during the annual cycle.

apparent temperature—A measure of the combined effect of high air temperature and high **relative humidity** on human comfort and well-being. The index currently used by the National Weather Service is based upon human physiology studies and is a measure of the ability of the human body to cool itself. It can be used to describe what hot weather "feels like" to the average person for various temperatures and relative humidities. Units of apparent temperature are in degrees Fahrenheit (or Celsius). Designed such that the apparent temperature is higher than the ambient temperatures for high humidities, and apparent temperature lower than the ambient temperature for low humidities. See the **heat index** or **temperature–humidity index (THI)**.

apparent wind—A **wind vector** relative to a moving object. The apparent **wind velocity** is the vector difference between the velocities of the true wind and the object.

applied climatology—The use of scientific analysis of climatic data for an operational purpose, such as for industrial, agricultural, or technical pursuits.

applied meteorology—The application of current weather data, analyses, and/or forecasts to specific practical problems. It is distinguished from **applied climatology**, which deals with the similar application of long-period statistically treated weather data.

aquifer—A geologic formation consisting of porous and permeable earth material (rock, sediment, soil) that contains and transmits large quantities of exploitable **groundwater**. The Ogalalla aquifer, extending from Texas to the Dakotas, is an example.

Archimedes' principle—The statement that a net upward or buoyant force, equal in magnitude to the weight of the displaced **fluid**, acts upon a body either partly or wholly submerged in a fluid at rest under the influence of **gravity**. Named for Archimedes (287–212 BC), a Greek mathematician who discovered the principle.

arctic air—A cold, dry **air mass** that forms over snow- and ice-covered surfaces of Siberia, the Arctic Basin, Greenland, and North America north of 60°N; responsible for the bitter **cold waves** that sweep across the Great Plains and Northeast of the United States.

Arctic Circle—The line of latitude 66° 33'N (often taken as 66.5°N). Along this line the sun does not set on the day of the boreal **summer solstice**, about 21 June, and does not rise on the day of the **winter solstice**, about 22 December. From this line, the number of 24-hour periods of continuous daylight or of continuous night increases northward to about six months each at the geographic North Pole.

Arctic Convergence—A zone in the major oceans of the Northern Hemisphere where **water masses** converge; it tends to be more poorly defined than its Southern Hemisphere counterpart, the **Antarctic Convergence**.

arctic desert—"Any area in the high latitudes dominated by bare rocks, ice, or snow, and having a sparse vegetation and a low annual precipitation." (*Glossary of Arctic and Subarctic Terms*, Arctic, Desert, Tropic Information Center Pub. A-105, 1955.) This includes portions of both **ice cap** and **tundra** regions of both hemispheres.

arctic front—The semipermanent, semicontinuous **front** between the cold **arctic air** to the north over the Arctic Basin and the less cold **polar air**; generally comparable to the **antarctic front** of the Southern Hemisphere.

arctic haze—A condition of reduced **horizontal** and slant **visibility** (but **unimpeded vertical visibility**) encountered by aircraft in flight (to above 9100 meters or 30 000 feet) over arctic regions. When viewed away from the sun it appears grayish-blue; into the sun it appears reddish-brown. It has no distinct upper and lower boundaries, and produces none of the optical phenomena that would be expected if it were composed of ice crystals. Color effects suggest particle sizes of two micrometers or less; generally believed to be composed of industrial pollution advected from lower latitudes, especially Europe.

arctic high (also known as *polar anticyclone*)—A weak **anticyclone** that appears on charts of mean sea level pressure over the Arctic Basin during late spring, summer, and early autumn.

arctic mist—A suspension in the air of **ice crystals**; a very light **ice fog**.

arctic sea smoke—Same as **steam fog**, but often specifically applied to steam fog rising from small areas of open water within **sea ice**.

arctic stratospheric vortex—Persistent westerly stratospheric circulation during winter in the Northern Hemisphere; it is most intense between 60° and 70°N and increases with altitude up to the **stratopause**. Though of jet stream strength, its maximum speeds are less than half those of the **antarctic stratospheric vortex**.

arctic tree line—The northern limit of tree growth; the sinuous boundary between **tundra** and **boreal forest**; taken by many to delineate the actual southern boundary of the **Arctic Zone** (2).

Arctic Zone—(1) Geographically, the area north of the **Arctic Circle** (66° 33'N). (2) Biogeographically, the area extending northward from the arctic **tree line** to the "limit of life." Same as **tundra**. It is also used for the level above the **timber line** in mountains.

arcus (also known as *arc cloud*)—A dense, arched-shaped, menacing-looking **accessory cloud** to a **cumulonimbus** that can occur along the leading edge of a thunderstorm's **gust front** as the consequence of uplift of stable warm air. Same as a **shelf cloud**.

area source—In air pollution studies, an extended origin of pollution from a series of points or lines spaced sufficiently close that the **plume** may be regarded as emanating from an area; an example is an industrialized city. Contrast with **line source** and **point source**.

Argon (Ar)—A colorless, odorless, inert **gas** that is the third most abundant gas species (by volume) in the **homosphere**, representing 0.934% by volume of **dry air**. Argon (atomic formula is Ar) has an atomic weight equaling 39.948 and is formed by natural radioactive decay of an unstable isotope of potassium. Often used as a tracer because of its inertness.

arid climate—A locale with a **climate** characterized by insufficient moisture to support appreciable plant life. See **desert climate**.

arid zone—(1) A region with insufficient moisture, where **evaporation** exceeds **precipitation**. (2) Sometimes used synonymously with **equatorial dry zone**.

aridity—The degree to which a climate lacks effective, life-promoting moisture. The opposite of **humidity** (1).

artesian water—**Groundwater** under sufficient pressure in an **aquifer** that is sandwiched between layers of impermeable rock or sediment.

artesian well—A shaft sunk into an **aquifer** containing **artesian water**. Because the **groundwater** is under pressure, it flows freely out of the well without requiring pumping.

artificial horizon—A planar reflecting surface that can be adjusted to coincide with the *astronomical horizon* (see **horizon**)—that is, can be made perpendicular to the **zenith**.

As—Abbreviation for the cloud type ***altostratus***.

Asiatic low—A **trough** of low atmospheric pressure located over South Asia, especially India, during the high sun season. It is a principal **center of action** in the atmospheric circulation of the Northern Hemisphere.

ASOS—Acronym for **Automated Surface Observing System**.

aspect—The direction toward which a land slope faces. The direction, measured relative to **true north**, is taken downslope and perpendicular to the **contours** of **elevation**.

aspirated psychrometer—An instrument for measuring atmospheric **humidity**, that utilizes a fan as an aspirator to force air past the **wet-bulb thermometer** for ventilation. Contrast with a **sling psychrometer**. See **Assmann psychrometer**.

aspirator—A device attached to a meteorological instrument to provide ventilation.

Assmann psychrometer—A fan-driven **aspirated psychrometer** with a special **radiation shield** of the thermal sensors. Designed by Richard Assmann (1845–1918), an aerologist.

astraphobia—A morbid fear of lightning and thunder.

astronomical refraction—(1) The angular difference between the apparent zenith distance of a celestial body and its true zenith distance, produced by **refraction** effects as the light from the body penetrates the atmosphere obliquely and is bent slightly. The apparent position typically is above its true position. The amount of refraction depends upon the **altitude** of the object and upon atmospheric conditions. (2) Any refraction phenomenon observed in the light originating from a source outside the earth's atmosphere. This applies only to refraction caused by inhomogeneities of the atmosphere itself, and not to that caused by ice crystals suspended in the atmosphere, as associated with **halo** phenomena.

astronomical scintillation—Any **scintillation** phenomena, such as irregular oscillatory motion, variation of intensity, or color fluctuation, observed in the light emanating from an extraterrestrial source; to be distinguished from **terrestrial scintillation** primarily in that the light source for the latter lies somewhere within the earth's atmosphere.

astronomical twilight—The interval of incomplete darkness before **sunrise** (or after **sunset**) when the true position of the center of the sun is between 12° and 18° below the **horizon**. At the time when the **solar zenith angle** is 108°, no discernible horizon glow appears over the sun's **azimuth**, and sixth magnitude stars can be seen near zenith.

atmosphere—(1) The envelope of air surrounding the planet earth and bound to it more or less permanently by virtue of the earth's gravitational attraction; the system whose chemical properties, dynamic motions, and physical processes constitute the subject matter of **meteorology**. Compare with **biosphere**, **cryosphere**, **hydrosphere**, and **lithosphere**. (2) As a unit of **atmospheric pressure**, see **standard atmosphere** (2).

atmospheric boil—Same as **terrestrial scintillation**.

atmospheric boundary layer—See **surface boundary layer** or **planetary boundary layer**.

atmospheric effect—See **greenhouse effect**.

atmospheric electric field—A quantitative term, denoting the strength of the **electric field** of the atmosphere at any specified point in space and time. In fair weather, this field may exceed 100 volts per meter near the earth's surface.

atmospheric electricity—(1) Electrical phenomena, regarded collectively, that occur in the earth's atmosphere. These phenomena include not only such striking manifestations as **lightning** and **Saint Elmo's fire**, but also less noticeable but more ubiquitous effects such as atmospheric **ionization** and other quiescent electrical processes. (2) The study of electrical processes occurring within the atmosphere.

atmospheric model—Description of the atmosphere giving a simplified or schematic outline of its key structures and processes via representative diagrams (conceptuals), systems of mathematical equations, or numerical approximations. The **standard atmosphere** is an example.

atmospheric optics—Study of the optical properties of the atmosphere and the optical phenomena or **photometeors** (such as **rainbows** and **halos**) produced by precipitation or the atmosphere's constituent **gases** and **aerosols**.

atmospheric phenomenon—As commonly used in weather observing practice, an observable occurrence of weather events of particular physical significance (e.g., **thunderstorms**, **tornadoes**, **waterspouts**) and **obstructions to vision**. Sometimes called "**weather**"(2).

atmospheric pollution—Contaminants present in the atmosphere, such as dust, gases, smoke, or water vapor in such quantities and with such characteristics and duration as to be injurious to human, plant, and animal life or property. These contaminants may be natural or anthropogenic.

atmospheric pressure—The force exerted per unit of area by the atmosphere as a consequence of gravitational attraction upon the "column" of air lying directly above the point in question. As with any **gas** the pressure exerted by the atmosphere is ultimately explainable in terms of bombardment by gas molecules; it is independent of the orientation of the surface on which it acts. Pressure is measured with a **barometer** or **barograph**. Units of atmospheric pressure are **hectopascals (millibars)** or millimeters (inches) of mercury.

atmospheric radiation—**Infrared radiation** (energy in the wavelength interval of 3–80 micrometers) emitted by or being propagated through the atmosphere; contains both **upwelling** and **downwelling** components. Compare with **terrestrial radiation**.

atmospheric shell—Any one of a number of strata or "layers" of the earth's atmosphere, as ascertained by criteria such as temperature, chemical composition, or ion concentration.

atmospheric shimmer—Same as **terrestrial scintillation**.

atmospheric tide—Analogous to ocean tides, periodic atmospheric fluctuations on the spatial scale of the planet earth, in which vertical **accelerations** are neglected (but compressibility is taken into account). Both the sun and moon produce atmospheric tides, through gravitational and thermal forcing.

atmospheric turbulence—Apparently random atmospheric motions; they are effective in transferring heat, moisture, and momentum but are so complex that only their statistical properties can be recognized and studied.

atmospheric wave—Generally, any pattern with some roughly identifiable periodicity in time and/or space. In meteorology, waves in the horizontal flow pattern (e.g., **Rossby wave**, **long wave**, **short wave**).

atmospheric window—Spectral range in which the atmospheric constituents absorb very little **electromagnetic radiation**. The visible window (0.3–0.8 micrometers) and the infrared window (8.5–11 micrometers) are notable examples.

atmospherics (also known as *sferics*)—Natural **electrical discharges** of a radio frequency that disturb radio communications. Atmospherics are heard as a quasi-steady background of crackling noise (static) in ordinary AM radio receivers. The discharges originate, principally, in the irregular surges of charge in **lightning discharges** of **thunderstorms**.

attenuation—In general, any decrease in amplitude, density, or energy as result of an effect such as **scattering**, **absorption**, or **friction**. In physical meteorology, a reduction in radiation flow, especially **solar radiation** by atmospheric gases and aerosols. In **radar meteorology**, the decrease in the magnitude of current, voltage, power, or intensity of a signal in transmission between points. Attenuation may be caused by interference such as rain or clouds.

audio frequency—The range of frequencies in **acoustic waves** that the normal human ear can hear; this ranges from 20 to 20 000 Hertz.

aureole—A poorly developed **corona** (1), characterized by a bluish-white disk immediately around the luminary and a reddish-brown outer edge; the radius is less than 5°, usually produced by **diffraction** of light passing through a collection of water droplets with a large size distribution.

aurora—A luminous phenomenon in the night sky consisting of overlapping curtains of greenish-white light sometimes fringed with pink appearing in the **ionosphere** of high latitudes. Aurorae are caused when the excited electrons of the atmospheric gases at altitudes of 100 kilometers or more recombine with the molecules, returning to their low energy states; exciting luminescence, analogous to neon tubes, results. The electrons are sent to higher energy levels or ejected from the molecules by high energy particles coming from the sun in the **solar wind**. These particles are ducted poleward by the earth's **magnetosphere**.

aurora australis—The "Southern Lights" or the **aurora** of the Southern Hemisphere.

aurora borealis—The "Northern Lights" or the **aurora** of the Northern Hemisphere.

auroral ovals—Roughly circular belts around either **geomagnetic pole** within which a maximum of auroral activity is located. These belts lie about 10°–15° of **geomagnetic latitude** from the geomagnetic poles.

autoconvection—The phenomenon of the spontaneous initiation of **convection** in an atmospheric layer in which the **environmental lapse rate** is equal to or greater than the **autoconvective lapse rate**.

autoconvective lapse rate—The **environmental lapse rate** of temperature in an atmosphere in which the mass density is constant with altitude. For dry air, the autoconvective lapse rate is approximately +34°C per kilometer. The lowest layers of the atmosphere immediately above a hot desert pavement may exceed this lapse rate.

Automated Surface Observing System (**ASOS**)—A widely employed, standardized set of automated weather instruments that provide routine **surface weather observations** of air **temperature**, **wind**, **dewpoint**, **visibility**, and **precipitation**.

autumn—(1) The **season** of the year that is the transition period from **summer** to **winter**, occurring as the sun approaches the **winter solstice**. In popular usage and for most meteorological purposes, autumn is customarily taken to include the months of September, October, and November in the Northern Hemisphere, and March, April, and May in the Southern Hemisphere. (2) Astronomically, the period extending from the **autumnal equinox** to the **winter solstice**.

autumn ice—**Sea ice** in its early stage of formation. It is comparatively salty, and crystalline in appearance, and not yet affected by lateral pressure.

autumnal—Pertaining to fall (**autumn**). The corresponding adjectives for **winter**, **spring**, and **summer** are **hibernal**, **vernal**, and **aestival**.

autumnal equinox—The **equinox** at which the sun approaches the Southern Hemisphere, marking the start of astronomical **autumn** in the Northern Hemisphere. The time of this occurrence is approximately 22 September. On that day, daylight is everywhere 12 hours. Compare with the **vernal equinox**, offset by six months.

available soil moisture—Water in the soil available to plants. Normally taken as the water in the soil between the **wilting point** and **field capacity**.

avalanche (also known as *snowslide*)—A mass of snow (perhaps containing ice and rocks) moving rapidly down a steep mountain slope and often taking with it earth, rocks, and rubble.

avalanche wind—The rush of air produced in front of an **avalanche** of dry snow or in front of a landslide.

aviation weather forecast—A forecast of **weather elements** of particular interest to aviation. These elements include the **ceiling**, **visibility**, **upper winds**, **aircraft icing**, **turbulence**, and types of precipitation and/or storms.

aviation weather observation—An evaluation, according to set procedure, of those **weather elements** that are most important for aircraft operations. It always includes **the cloud height** or **vertical visibility**, **sky cover**, **visibility**, **obstructions to vision**, certain **atmospheric phenomena**, and **wind speed** and **direction** that prevail at the time of the observation. Complete observations include the **sea level pressure**, temperature, dewpoint temperature, and **altimeter setting**. These observations are encoded and transmitted over electronic networks for use by the aviation and meteorological community. See **record** and **special observations**.

Avogadro's number—The number of molecules in one **mole** of **gas** (6.022169×10^{23} per mole). According to *Avogadro's law,* this number is a constant for permanent gases under normal conditions—that is, pressure of one **standard atmosphere** and temperature of 0°C (32°F)—the volume occupied by one mole of gas is the same for all permanent gases (22 421 cubic centimeters or 22.42 liters). Named for Amedo Avogadro (1776–1856), an Italian chemist who identified this relationship.

AWIPS—Acronym for **Advanced Weather Interactive Processing System**.

azimuth or **azimuth angle**—The length of arc measured clockwise along the astronomical **horizon** (in degrees of arc) from the adopted reference direction, usually **true north**, to that point on the horizon where the particular object or its projection is located; north is defined as 0° (or 360°), east is 90°, and so forth. See **bearing**.

Azores high—The semipermanent atmospheric **subtropical anticyclone** over the North Atlantic Ocean, so named especially when it is located over the eastern part of the ocean, near the Azores Islands. In summer, the axis of this high is located at 35°N, with a **central pressure** of 1024 hectopascals (millibars); in winter, it has shifted to 30°N, with little change in central pressure

B

back-door cold front—A **cold front** that leads a cold **air mass** toward the south and southwest along the Atlantic seaboard of the United States; so named because of the somewhat unusual movement of the front.

background luminance—The **brightness** of the background against which a target is viewed, especially important in identifying range markers.

backing—A change in **wind direction** in a counterclockwise sense (e.g., south to southeast to east) in the Northern Hemisphere of the earth. Opposite of **veering**.

backlash—A region of precipitation and unsettled weather related to the circulation from around the north and west sides that trail around a Northern Hemisphere migratory storm.

back-scattering—Same as **backward scatter**.

backward scatter—The **scattering** of **radiant energy** into the hemisphere of space bounded by a plane normal to the direction of the incident radiation and lying on the same side as the incident ray. Opposite of **forward scatter**. Atmospheric backward scatter depletes 6%–9% of the incident solar beam before it reaches the earth's surface. In radar usage, backward scatter refers only to that radiation scattered at 180° to the direction of the incident wave, back toward the radar antenna to produce an **echo** on the **radar screen**.

backwash—(1) Intermittent seaward flow of water across a beach; the return of **swash** to the sea. (2) Reversal of the water flow under pressure.

ball lightning (also called *globe lightning*)—A relatively rare form of **lightning**, consisting of a (usually) reddish, luminous ball, of the order of 30 centimeters in diameter, which may move rapidly along solid objects or remain floating in mid-air. Hissing noises emanate from such balls, and they sometimes explode noisily, but may also disappear noiselessly.

balloon drag—A small balloon, loaded with ballast and inflated so that it will explode at a predetermined altitude, that is attached to a larger balloon. It is frequently used to retard the ascent of a **radiosonde** during the early part of the flight, so that more detailed measurements may be obtained.

band lightning—Same as **ribbon lightning**.

banded structure—The appearance of **precipitation echoes** in the form of long bands as presented on **radar reflectivity** displays.

bandwidth—The range of **frequencies (Hertz)** between the limits of a frequency band. Bandwidth is a measure of the faithfulness with which radio energy input is passed through the receiver without distortion or loss of data. It is one of the variables determining the minimum detectable signal of a **radar** unit. The shorter the pulse duration, the larger the bandwidth required to preserve the same quality of receiver output pulses.

bankfull stage—The **stage**, on a fixed **river gauge**, corresponding to the top of the lowest banks within the **reach** for which the gauge is used as an index.

banner cloud—A stationary, orographically produced cloud plume often observed to extend downwind from isolated mountain peaks, even on otherwise cloud-free days. See **standing cloud**.

bar—(1) A unit of **pressure** (force per unit area) equal to 100 000 newtons per square meter, 1000 millibars, 1000 hectopascals, or 29.53 inches of mercury. (2) An offshore ridge or mound that is submerged for some portion of the **tide cycle**. It is often found at the mouth of a river or an **estuary**. See **shoal**.

barb—A means of representing **wind speed** in the plotting of weather information on a **synoptic chart**; it is a short straight line drawn obliquely toward lower pressure from the end of a **wind direction shaft**. In the commonly used five-and-ten system, one barb represents a wind speed of 10 knots and a half-barb represents 5 knots.

barber—(1) A severe storm at sea during which spray and precipitation freeze onto the decks and rigging of boats. (2) In the Gulf of St. Lawrence, a local form of **blizzard** in which wind-borne ice particles almost cut the skin from the face. (3) A rare type of **fog** formed in the same manner as **steam fog** but at colder temperatures so that it is composed of ice particles instead of water droplets.

barchan—A crescent-shaped dune or drift of wind-blown sand or snow; the arms of the crescent point downwind.

baroclinic (or *baroclinic atmosphere*)—An atmosphere in which temperature (and hence density) gradients exist upon **constant pressure surfaces**; an atmosphere that is not **barotropic**. The atmospheric region in the vicinity of a **cold front** is an extreme example. In a baroclinic atmosphere, the **geostrophic wind** varies with height.

baroclinic disturbance—Any **migratory cyclone** more or less associated with strong **baroclinity** of the atmosphere, evidenced on synoptic weather charts by **temperature gradients** in the constant-pressure surfaces, vertical **wind shear**, and **tilt** of pressure troughs with height.

baroclinic instability—A **hydrodynamic instability** in a **fluid** (such as the atmosphere) characterized by a horizontal **temperature gradient** and hence a vertical **shear** in the flow. This instability results in a conversion of the available **potential energy** of the mean flow into the **kinetic energy** of the **baroclinic disturbance**.

baroclinity (or *baroclinicity*)—A measure of the state of stratification in a **fluid** in which surfaces of constant pressure (**isobaric**) intersect surfaces of constant density (**isosteric**).

barograph—A recording instrument that provides a continuous trace of **air pressure** variation with time.

barometer—An instrument for measuring **atmospheric pressure**. Two types of barometers are commonly used in meteorology: the **mercury barometer** and the **aneroid barometer**.

barometric corrections—Systematic adjustments that must be applied to the reading of **a mercury barometer** in order that observed values may be accurate or standardized. Four kinds of corrections are made: (*a*) The *instrument correction* is the mean difference between the readings of a given mercury barometer and those of a standard instrument. It is a composite correction, including the effects of capillarity (see **capillarity correction**), index misalignment, imperfect vacuum, and scale correction. (*b*) The *temperature correction* is applied to adjust the scale reading made at the temperature of the barometer to standard temperature, accounting for the difference between the coefficient of expansion of mercury and that of the scale. (*c*) The *gravity correction* is applied to adjust the local **gravity** to **standard gravity**, incorporating the altitude and latitude of the station; this correction is necessary because the acceleration of gravity varies both with altitude and latitude. (*d*) The *removal correction* is applied when the barometer elevation differs from the adopted **station elevation**.

barometric gradient—Same as **pressure gradient**.

barometric hypsometry—The technique of estimating **elevation** by means of atmospheric pressure measurements.

barometric pressure—Same as **atmospheric pressure**.

barometric tendency—Same as **pressure tendency**.

barometric wave—Any wavelike variation in the atmospheric pressure field. The term is usually reserved for short-period variations *not* associated with **synoptic-scale** motions or with **atmospheric tides**.

barothermograph—An instrument that automatically and simultaneously records pressure and temperature on one chart.

barotropic—Of, pertaining to, or characterized by a condition of **barotropy**; surfaces of constant density parallel surfaces of constant pressure. Contrast with **baroclinic**.

barotropic instability—The **hydrodynamic instability** arising from certain distributions of **vorticity** in a two-dimensional nondivergent flow.

barotropic model—Any of a number of **model atmospheres** in which some of the following conditions exist throughout the motion: coincidence of pressure and temperature surfaces, absence of vertical **geostrophic wind shear**, absence of vertical motions, absence of horizontal velocity **divergence**, and conservation of the vertical component of **absolute vorticity**.

barotropy—The state of a **fluid** in which surfaces of constant density (or temperature) are coincident with surfaces of constant pressure; it is the state of zero **baroclinity**.

barrens—Any region that is devoid of vegetation or permits only the sparse growth of very few plant species. This term is most commonly applied to such terrain in polar regions. See **arctic desert**.

barrier jet—A strong air current induced by **orographic** effects; often produced when a cold air mass surges against a prominent mountain range.

base map—A map designed for the presentation and analysis of data; it usually includes only the coordinates, geographical and major political outlines, and sometimes the larger lakes and rivers.

basin—(1) See **river basin** or **drainage basin**. (2) See **ocean basin**, a large circular or oval-shaped depression on the ocean floor.

basin lag—A computed characteristic of a particular **river basin**, expressed as the time difference between the time-center of mass of **rainfall** and the time-center of mass of resulting **runoff**.

basin recharge—The difference between amounts of **precipitation** and **runoff** for a given storm; it is that portion of the precipitation that remains in the basin as **soil moisture**, **surface storage**, **groundwater**, etc.

bathymetric chart—A map delineating the form of the bottom of a body of water, usually by means of depth contours (**isobaths**).

bathythermograph (BT)—A device for obtaining a record of temperature against depth (strictly speaking, pressure) in the upper 300 meters of the ocean, from a ship under way. Some of these devices are expendable and designated as *XBT*.

bead lightning (also known as *pearl lightning*)—Merely a particular aspect of a normal **lightning flash** occasionally seen when the observer happens to view end-on a number of segments of the irregular channel (*zigzag lightning*) and hence receives an impression of higher intensity at a series of locations along the channel. Lightning resembles a series of beads tied to a string.

beam radiation—Same as **direct solar radiation**.

bearing—The horizontal direction from one terrestrial point to another; basically synonymous with **azimuth**. Bearing, however, may be expressed in several ways: *true bearing* and *magnetic bearing* are the angular directions in degrees measured clockwise from **true north** and **magnetic north**, respectively; *compass bearing* is expressed in terms of compass points; and *relative bearing* is the angular distance (in degrees) measured clockwise from the heading of a craft.

beating—A wave phenomenon that occurs when two or more waves of different frequencies become superimposed. The resultant wave has amplitude maxima (''beats'') at the frequency equal to the difference in the frequencies of the initial waves.

Beaufort wind scale—A system of estimating and reporting **wind speeds** invented in the early nineteenth century by Admiral Francis Beaufort (1774–1857) of the British Navy in 1805. It was originally based on the effects of various wind speeds on the amount of canvas that a full-rigged frigate of the period could carry, but has since been modified and modernized. In its present form for international meteorological use it equates (*a*) Beaufort force (or *Beaufort number* from 0 to 12), (*b*) wind speed, (*c*) descriptive term, and (*d*) visible effects upon land objects or sea surface (**state of the sea**).

Beer's law—In meteorology, same as **Bouguer's law**.

Benard cell—A form of **cellular convection**, usually hexagonal as viewed from above (when the **fluid** has a free surface), formed in thin layers of a fluid initially at rest and heated from below. When seen from a satellite perspective, some cloud patterns appear to have a similar cellular structure. Named for Henri Bénard (1874–1939), a French physicist and meteorologist who studied atmospheric motions.

Benguela Current—The cold northward-flowing ocean current in the South Atlantic, flowing along the west coast of Africa; it is one of the swiftest of ocean currents, and is the strongest current in the South Atlantic. It is a continuation of the **South Atlantic Current**. Proceeding toward the equator, the Benguela Current gradually leaves the coast and continues as the northern portion of the **South Equatorial Current**.

bent-back occlusion—Rotation of an **occluded front** in a counterclockwise sense (in the Northern Hemisphere) about a center of **low pressure** so that the front begins to progress toward the west.

berg—Commonly used abbreviation for **iceberg**.

Bergeron–Findeisen theory (also known as *ice crystal theory*)—A theoretical explanation of the process by which precipitation particles may form within a cloud composed of both ice crystals and supercooled liquid water droplets. This theory is based on the fact that the **equilibrium vapor pressure** of water vapor with respect to ice is less than that with respect to liquid water at the same subfreezing temperature. Thus, within a mixture of these particles, and provided that the total water content is sufficiently high, the ice crystals would gain mass by **deposition** at the expense of the liquid droplets, which would lose mass by **evaporation**.

Bermuda high—The semipermanent atmospheric **subtropical anticyclone** over the North Atlantic Ocean, so named especially when it is located in the western part of the ocean, near Bermuda (near 30° N). A westward extension of the **Azores high**.

Bernoulli's principle—Air flowing over an airfoil results in an increase in flow speed over the upper curved surface. Since a velocity increase in **fluid** flow results in a corresponding pressure decrease, the increased airflow over the upper surface of the airfoil produces a lift on the airfoil because of lower pressure exerted on the upper surface. Named for Daniel Bernoulli (1700–1782), a Swiss physicist who discovered the effect.

beta particle—A high-speed electron, ejected during the decay of certain radioactive substances. Compare with **alpha particle** and **gamma rays**.

biennial wind oscillation (also known as *quasi-biennial oscillation* or *QBO*)—A periodic alternation of easterly and westerly wind regimes in the **stratosphere**, within about 12° of the equator, with a period varying from 24 to 30 months.

billow cloud—Broad, nearly parallel lines of cloud oriented perpendicular to the **wind direction**, with cloud bases near a **temperature inversion** surface. The distance between billows is usually of the order of 1000–2000 meters.

bimetallic thermometer—A **deformation-type thermometer** or a thermal sensor with a sensing element consisting of a compound strip of metal formed by welding together two strips of metal having different coefficients of expansion. The curvature of the strip is a function of its temperature. Often used in **thermographs**.

biofog—Basically, a type of **steam fog** caused by contact between extremely cold air and the warm, **humid air** surrounding human or animal bodies or generated by human activity.

biogeochemical cycle—Circulatory paths or movements through the planet earth system of those key chemical constituents essential to life. These constituents include carbon, nitrogen, oxygen, and phosphorus, and each is identified with a named cycle, for example, **carbon cycle**. These cycles often include the **atmosphere** and **hydrosphere**.

bioluminescence—The production of light by a living organism as a result of biochemical reactions without **sensible heat**. These light producing reactions can take place either within the cells of the organism or as a secretion.

biomass—The total amount of living material in a given system or body of water, expressed as a mass (or weight).

biome—A well-defined environment or community describing the complex of living organisms found in that ecological region. Examples include **tundra**, **boreal forest**.

biometeorology—Study of the influences exerted on living organisms by the **weather elements** (e.g., temperature, humidity, precipitation).

biosphere—That transition zone between the earth and its atmosphere within which most forms of terrestrial life are commonly found; the outer portion of the **lithosphere** and inner or lower portion of the **atmosphere**.

bi-rainy climate—A climate characterized by the regular recurrence of two distinct **rainy seasons** annually. This is found in land regions near the equator where the heaviest rains occur shortly after the **equinoxes** (the **equinoctial rains**). It may be considered a subdivision of the **tropical climate**.

Bishop wave—A striking example of an atmospheric **lee wave**, formed to the lee of the Sierra Nevada range near Bishop, California. The phenomenon includes a **rotor cloud** and a series of **lenticular clouds** parallel to the crest of the range.

Bishop's ring—A faint, broad, reddish-brown **corona**-like phenomenon occasionally seen in dust clouds, especially those resulting from violent volcanic eruptions. The angular radius of this ring's inner edge is about 20° and its angular width is about 10°, the exact dimensions being controlled by the particle size of the dust layer. The existence of this ring is due to the **diffraction** of sunlight by the dust particles. It is named after Rev. S. Bishop of Honolulu, who first described the phenomenon following the 1883 Krakatau eruption.

black frost—A dry **freeze**, with respect to its effects upon vegetation—that is, the internal freezing of vegetation unaccompanied by the protective formation of **hoarfrost**. A black frost is always a *killing frost*, and its name derives from the resulting blackened appearance of affected vegetation.

black ice—Thin, new ice on **fresh-** or **saltwater**, appearing dark in color because of its transparency; also popularly applied to thin hazardous ice coverings on roads.

blackbody—A hypothetical "body" that absorbs all of the **electromagnetic radiation** striking it—that is, one that neither reflects nor transmits any of the incident radiation. The radiation **emittance** is consistent with **Planck's law**. In accordance with **Kirchhoff's law**, a blackbody not only absorbs all wavelengths, but emits at all wavelengths and does so with maximum possible intensity for any given temperature. Contrast with **whitebody** and **graybody**.

blackbody radiation—The **electromagnetic radiation** emitted by an ideal **blackbody** adhering to the **radiation laws**; it is the theoretical maximum amount of **electromagnetic radiation** of all wavelengths that can be emitted by a body at a given temperature.

blind drainage—Same as **closed drainage**.

blind rollers—Long **swells** whose height increases to almost the breaking point as they pass over **shoals** or run in shallow water.

blizzard—A severe weather condition lasting at least 3 hours, characterized by low temperatures and by strong winds bearing a great amount of **blowing snow**, which reduces the visibility to less than 1 kilometer. The National Weather Service specifies, for *blizzard*, a wind of 35 mph or higher, low temperatures, and sufficient snow in the air to reduce **visibility** to less than 0.25 mile (400 meters). The name originated in the United States (possibly in Virginia) but it has spread to similar winds in other countries. In popular usage in the United States and in England, the term is often used for any heavy snowstorm accompanied by strong winds. Although there is no set temperature requirement for blizzard conditions, the life-threatening nature of the low temperature in combination with the other hazardous conditions of wind, snow, and poor visibility increases dramatically when the temperature falls below 20°F (–7°C).

blizzard warning—Issued by the National Weather Service to warn the public of a potentially life threatening winter storm condition with sustained winds or frequent **gusts** to 35 mph or higher and considerable falling and/or **blowing snow,** which reduces **visibility** to less than 0.25 mile (400 meters). These conditions are expected to last for at least 3 hours.

blocking—The obstructing, on a large scale, of the normal west-to-east progression of **cyclones** and **anticyclones** in exaggerated high-amplitude **meridional flow**. See **cutting-off process** and **blocking high**.

blocking high—Any **anticyclone** that remains nearly stationary or moves slowly compared to the west-to-east motion "upstream" from its location, so that it effectively "blocks" the movement of migratory **cyclones** across its latitudes. See **omega block**.

blood rain—Rain of reddish color caused by foreign matter picked up by raindrops during descent. A dust-filled subcloud layer is required to yield this effect, and the dust particles must contain sufficient iron oxide to be red in color.

blowing dust—A **lithometeor** phenomenon of **dust** lifted from the earth's surface and blown about locally to considerable heights by the wind. In United States observational procedure, blowing dust is reported if the amount of suspended dust reduces **horizontal visibility** to 6 miles (9 kilometers) or less. In the extreme, a **duststorm**.

blowing dust/blowing sand advisory—Issued by the National Weather Service to inform the public that **blowing dust** or **blowing sand** will reduce visibility to 0.25 mile (0.4 kilometer) or less. Contrast with **duststorm warning**.

blowing sand—A **lithometeor** phenomenon of sand grains raised from the earth's surface and blown about locally to considerable heights by the wind. In United States observational procedure, blowing sand is reported if the amount of suspended sand reduces **horizontal visibility** to 6 miles (9 kilometers) or less. In the extreme, a **sandstorm**.

blowing snow—Loose snow lifted from the earth's surface by the wind to an eye-level height of six feet or more above the surface (higher than **drifting snow**) and blown about in such quantities that **horizontal visibility** is restricted at and above that height. Blowing snow is one of the classic requirements for a **blizzard**.

blowing snow advisory—Issued by the National Weather Service to alert the public of the possibility that wind-driven snow (the condition of **blowing snow**) will reduce **visibility** enough to hamper travel. Visibility conditions are not expected to deteriorate sufficiently to issue a **blizzard warning**.

blowing spray—**Spray** lifted from the sea surface by the wind and blown about in such quantities that the **horizontal visibility** is restricted.

blue ice—Pure ice in the form of large single crystals. The blue color is caused by the wavelength-selective **scattering** of **polychromatic** light by the ice molecules; the purer the ice, the deeper the blue.

blue moon—(1) A rare phenomenon caused by the presence of large quantities of suspended particles in the atmosphere that selectively remove the longer reddish-colored wavelengths of the sunlight reflected from the lunar surface more than the blue or green wavelengths. This event has been observed after large forest fires. (2) A name given to the second full moon in a given calendar month; occurs approximately once every three years.

boiling point (also known as the *steam point*)—The temperature at which the **equilibrium vapor pressure** between a **liquid** and its **vapor** is equal to the external pressure on the liquid. Because of the normal decrease of **barometric pressure** with altitude, the nominal boiling point of water decreases 3.3°C per 1000 meters (1.8°F per 1000 feet) increase of altitude. The boiling point is a colligative property of a solution; with an increase in dissolved matter, a raising of the boiling point occurs. The boiling point of pure water at standard pressure is equal to 100°C (212°F).

Boltzmann's constant—The ratio of the universal **gas constant** to **Avogadro's number**; equal to 1.38062×10^{-23} joules per kelvin. Named for Ludwig Boltzmann (1844–1906), an Austrian physicist.

bomb—A **cyclone** exhibiting explosive **deepening** averaging 1 hectopascal (1 millibar) per hour for 24 hours. Such systems result from an **upper-level trough** coinciding with a strong **gradient** of **sea surface temperature**.

bora—A **fall wind**, or a **gravity**-driven downslope wind whose source is so cold that when the air reaches the lowlands or coast, compressional warming is insufficient to raise the air temperature to the normal level for the region; hence, it appears as a cold wind. The terms *borino* and *boraccia* denote a weak bora and strong bora, respectively. The namesake wind regime occurs along the Dalmatian coast (Croatia) in winter.

bora fog—A dense ''fog'' caused when the **bora** lifts a spray of small drops from the surface of the sea.

bore—See **tidal bore**.

boreal forest—The forested region that adjoins the **tundra** along the **arctic tree line**. It has two main divisions: its northern portion is a belt of **taiga** or *boreal woodland*; its southern portion is a belt of true forest, mainly conifers but with some hardwoods. On its southern boundary the boreal forest passes into ''mixed forest'' or ''parkland,'' **prairie** or **steppe**, depending on the rainfall.

Boreal zone—A biogeographical region characterized by a northern type of fauna or flora; includes the area of North America situated between the mean summer **isotherm** of 18°C (64.4°F) (roughly 45°N) and the **Arctic Zone**.

Boreas—The ancient Greek name for the north wind (now also *borras*). Being cold and stormy, it is represented on the **Tower of the Winds** in Athens by a warmly clad old man carrying a conch shell (probably to represent the howling of the wind). The term may originally have meant ''wind from the mountains,'' thus, the present use of **bora**.

bottom water—The **water mass** at the deepest part of an ocean water column. It is the densest water that is permitted to occupy that position by the regional topography. In the case of an **ocean basin**, bottom water may be formed locally, or it may represent the densest water that has existed at **sill** depth in the recent past.

Bouguer's law (*Beer's law* or *Lambert's law*)—A relationship describing the rate of decrease of intensity of a plane-parallel beam of **monochromatic** (single wavelength) radiation as it penetrates a medium that both scatters and absorbs at that wavelength; expressed as an exponential decay relationship. Pierre Bouguer (1698–1758), a French mathematician; August Beer (1825–1863), a German physicist; and Johann Heinrich Lambert (1728–1777), a German mathematician and astronomer, formulated this relationship.

boundary conditions—Stipulations or circumstances pertaining to conditions at the edges or physical boundaries of a domain that must be satisfied in solving a problem.

boundary layer—The layer of **fluid** in the immediate vicinity of a bounding surface, usually in reference to the **planetary boundary layer** or **surface boundary layer**.

Bowen ratio—For any moist surface, the ratio of **heat** energy used for **sensible heating** (conduction and convection) to the heat energy used for **latent heating** (evaporation of water or sublimation of snow). Bowen ratio ranges from about 0.1 for the ocean surface to more than 2.0 for deserts; negative values are also possible. It is named for Ira S. Bowen (1898–1978), an American astrophysicist.

Boyle's law—The empirical generalization that for many so-called **perfect gases**, the product of pressure and volume is constant in an **isothermal** process. Named for Robert Boyle (1627–1691), a British chemist who formulated this relationship.

Brazil Current—The warm ocean current in the South Atlantic Ocean flowing southward along the Brazilian coast. Its origin is in the westward-flowing **South Equatorial Current**, part of which turns south and flows along the South American coast as the Brazil Current, a tongue of water of relatively high temperature and high **salinity**. At about 35°S it meets the **Falkland Current**, where the two turn east and cross the ocean as the **South Atlantic Current**.

breaker—A sea **surface wave** that has become too steep to be stable, especially in shallow or shoaling water.

breaks in overcast—In United States weather observing practice, a condition wherein the **sky (cloud) cover** is more that 0.9 but less than 1.0 (to the nearest tenth). This would appear in an **aviation weather observation** as an **overcast** sky cover.

breakup—In general, the spring melting of snow, ice, and frozen ground. Specifically, the destruction of the **ice cover** on rivers during the spring thaw; or applied to the time when the solid sheet of ice on rivers breaks into pieces that move with the current. A phenological event in northern latitudes.

breeze—(1) In general, a **light wind**. (2) In the **Beaufort wind scale** (*Beaufort force numbers 2–6*), a **wind speed** ranging from 4 to 27 knots (4 to 31 mph) and categorized as follows: *light breeze*, 4–6 knots; *gentle breeze*, 7–10 knots; *moderate breeze*, 11–16 knots; *fresh breeze*, 17–21 knots; and *strong breeze*, 22–27 knots.

bright band—The narrow enhanced intensity in a **radar reflectivity** pattern of snow as it melts to rain, as displayed on a **range-height indicator** scope; results from water coated ice particles exhibiting a higher **reflectivity** at the **melting level**.

bright segment (also called ***twilight arch***)—A faintly glowing band that appears above the horizon after sunset or before sunrise; its disappearance after sunset marks the end of **astronomical twilight** and the beginning of full darkness; a **twilight phenomenon**.

brightness—A basic visual sensation describing the amount of light that appears to emanate from an object, or more precisely, the **luminance** of an object.

brightness temperature—The apparent temperature of a celestial object, based on the assumption that it radiates as a **blackbody**.

brine—**Seawater** containing a higher concentration of dissolved salt than the usual **salinity** of ocean water; often produced by evaporation or freezing of ocean water.

British thermal unit (**BTU** or **Btu**)—A unit of **energy** defined as the heat required to raise the temperature of one pound of water one Fahrenheit degree at 60°F; it is equal to 1055 joules (252.1 calories).

broken—An official **sky cover** classification for **aviation weather observations**, descriptive of a sky cover of from 0.6 to 0.9 (to the nearest tenth). This is applied only when **obscuring phenomena** aloft are present—that is, not when obscuring phenomena are surface-based, such as **fog**.

brown snow—Snow intermixed with dust particles; a not uncommon phenomenon in many parts of the world. Snows of other colors, such as **yellow snow**, are similarly explainable.

Brownian motion—The incessant, random movements exhibited by the dispersed particles in a **colloidal system**, resulting from random collisions between the molecules of the dispersing medium and the particles of the dispersed phase. Named for Robert Brown (1773–1858), a Scottish physician and botanist who first described this motion.

BT—Abbreviation for **bathythermograph**.

bubble high—A small **anticyclone** of the order of 80–500 kilometers (50–300 miles) across, often induced by precipitation and vertical currents associated with **thunderstorms**. These transitory small highs are relatively cold and have the effect of a different **air mass**; unstable air overrunning these bubbles may form **squall lines** at their boundary.

bubbly ice—**Glacier ice** containing air bubbles trapped when the ice was compressed in the **accumulation zone**.

bucket temperature—The surface temperature of the sea as measured by a **bucket thermometer** or by immersing any thermometer in a freshly drawn bucket of **seawater**.

bucket thermometer—In oceanography, a thermometer used in a bucket of **seawater** to measure **sea surface temperature**.

budget year—The one-year period beginning with the start of the **accumulation** season at the **firn line** of a **glacier** or ice cap and extending through the following summer's **ablation** season.

buoyancy—The upward force exerted upon a parcel of **fluid** (or an object within the fluid) in a gravitational field by virtue of the density difference between the parcel (or object) and that of the surrounding fluid.

buoyant force—The upward **force** exerted upon an object (e.g., **air parcel**) in a **fluid** in a gravitational field as a consequence of a density difference between the object and the ambient or environmental fluid. A positive buoyant force means an object less dense than its environment accelerates upward.

burga—A storm with northeasterly winds in Alaska, bringing **ice pellets** (**sleet**) or snow.

burn off—With reference to **fog** or low **stratus** cloud layers, to dissipate by daytime heating from the sun.

Buys Ballot's law—An empirical law describing the relationship of the horizontal **wind direction** in the atmosphere to the horizontal pressure distribution; with your back to the wind, the air pressure is lower to the left than to the right in the Northern Hemisphere. In the Southern Hemisphere, the relationship is reversed. Named for Christoph H.D. Buys-Ballot (1817–1890), a Dutch meteorologist who described this relationship.

C

calibration—(1) A measured comparison with a standard. (2) The process whereby a position on the scale of an instrument is identified with the magnitude of the signal actuating that instrument.

California Current—The cold ocean current flowing southward along the west coast of the United States from approximately Washington State to northern Baja California. It is the major branch of the **Aleutian Current**. As a whole, the current represents a wide body of water that moves sluggishly toward the southeast. Off Central America, the California current turns toward the west and becomes the **North Equatorial Current**.

calm—The absence of apparent motion of the air. In the **Beaufort wind scale**, this condition (*Beaufort force number 0*) is reported when smoke is observed to rise vertically, or the surface of the sea is smooth and mirrorlike. In United States weather observing practice, the wind is reported as calm if **wind speeds** of less than one mile per hour (or one knot) are observed.

calms of Cancer—Along with the ''calms of Capricorn,'' the light variable winds and calms that occur in the centers of the **subtropical high-pressure belts** over the oceans. They are named after the **Tropics of Cancer** and **Capricorn**, although their usual position is at about 30°N and 30°S; the **horse latitudes**.

calorie (cal)—A unit of **heat** defined as the amount of heat required to raise the temperature of one gram of water through one Celsius degree from 14.5°C to 15.5°C (the *gram-calorie* or *small calorie*). One calorie equals 4.1855 joule. The large calorie (Cal) used for dietary purposes is 1000 times larger than one small calorie.

calorimeter—An instrument designed to measure quantities of **heat**; sometimes used in meteorology to measure **solar radiation**.

calving—The breaking off of an ice mass from its parent **glacier**, **iceberg**, or **ice shelf**.

Canaries Current—The warm southern branch of the **North Atlantic Current** (which divides on the eastern side of the ocean); it moves south past Spain, the Canary Islands, and North Africa to join the **North Equatorial Current**.

candle ice—Disintegrating **sea ice** (or lake ice) consisting of ice prisms or cylinders oriented perpendicular to the original ice surface; these ''ice fingers'' may be equal in length to the thickness of the original ice before its disintegration.

canyon wind—A **local wind** regime described as the **mountain wind** of a canyon—that is, the nighttime downslope flow of air caused by cooling along the canyon walls. Because of the steepness of the slopes, canyon winds can be very strong.

cap cloud—An approximately stationary cloud, or **standing cloud**, on or hovering above an isolated mountain peak. It is formed by the cooling and condensation of **humid air** forced up over the peak. See **pileus**.

capacity—(1) Maximum volume of a substance (typically water) that can be contained in a **reservoir**. (2) Maximum flow rate (**flux**) that can be carried by any conveying structure.

capillarity correction—As applied to a **mercury barometer**, that part of the **instrument correction** that is required by the shape of the **meniscus** of the mercury. Mercury does not wet glass, and consequently the shape of the meniscus is normally convex upward, resulting in a positive correction. For a given barometer, this correction will vary slightly with the height of the meniscus. The capillarity correction can be minimized by using a tube of large bore.

capillary action—The depression or elevation of the **meniscus** of a **liquid** contained in a tube of small diameter due to the combined effects of **surface tension** and the attractive forces between the same type molecules(*cohesion*) and between molecules of different materials (*adhesion*). When the liquid wets the wall of a container, the meniscus is shaped convex downward; if the liquid does not wet the walls of the container, the meniscus is shaped convex upward.

capillary fringe (or *capillary zone*)—A shallow zone of permeable rock or soil above a **water table** containing water lifted from the water table by **capillary action**; the interstices in the soil are filled with water that is under less than **standard atmospheric pressure**.

capillary water—Water stored in the soil in the form of tiny droplets that are held together by **surface tension** or the molecular attraction between water and soil particles; typically found in the **capillary fringe**.

capillary wave (also called *ripple*)—A wave, on a **fluid** interface, of sufficiently short wavelength (less than 2 centimeters for water waves) that the primary restoring force is **surface tension**; smaller than a **gravity wave**.

captive balloon sounding—Measurement of one or more upper-air **weather elements** by means of a moored (or tethered) balloon carrying measuring instruments; especially used for studies in **micro**- and **mesometeorology**.

carbon cycle—One of the major **biogeochemical cycles** that operate on planet earth, consisting of the continuous movement of carbon in its many chemical forms from one reservoir to another; reservoirs in the carbon cycle include the **atmosphere**, **lithosphere** (carbonaceous rocks and sediments), **biosphere**, and the **hydrosphere** (ocean). The exchanges of carbon between reservoirs are governed by various physical, chemical, and biological processes, as well as by human intervention. The greatest reservoir of carbon appears to be the upper portion of the ocean, with large quantities of dissolved carbon dioxide (CO_2) and organic material.

carbon dioxide—A heavy, colorless **gas** of chemical formula CO_2, with a molecular weight of 44.010. It is the fourth most abundant constituent of **dry air** in the **homosphere**, now present to the extent of 0.035% by volume and considered to be a

greenhouse gas. Over 99% of the terrestrial CO_2 is found in the oceans, but its solubility is strongly temperature dependent, so changes in **sea surface temperatures** can lead to marked local changes in CO_2 content of the ocean.

cardinal winds—Winds from the four cardinal points of the compass—that is, north, east, south, and west winds.

Caribbean Current—An ocean current flowing westward through the Caribbean Sea. It is formed by the commingling of part of the waters of the **North Equatorial Current** with those of the **Guiana Current**. It flows through the Caribbean Sea as a strong current and continues with increased speed through the Yucatan Channel; there it bends sharply to the right and flows eastward with great speed out through the Straits of Florida to form the **Florida Current**.

Carnot cycle—An idealized reversible **work** cycle defined for any system, but usually limited, in meteorology, to a so-called **perfect gas**. The Carnot cycle consists of four stages: (*a*) an **isothermal** expansion of the gas at temperature T_1; (*b*) an **adiabatic** expansion to temperature T_2; (*c*) an isothermal compression at temperature T_2; (*d*) an adiabatic compression to the original state of the gas to complete the cycle. In a Carnot cycle, the net work done is the difference between the heat input Q_1 at higher temperature T_1 and the heat extracted Q_2 at the lower temperature T_2. Sadi Carnot (1796–1832), a French physicist, helped found thermodynamics theory.

Cartesian coordinates—A **coordinate system** named for René Descartes (1596–1650), the French mathematician, in which the locations of points in space are expressed by reference to three orthogonal planes, called *coordinate planes*, no two of which are parallel. The three planes intersect in three straight lines, called *coordinate axes*, commonly called x,y,z. The coordinate planes and coordinate axes intersect at a common point called the *origin*.

castellanus—A **cloud species** of which at least a fraction of its upper part presents some vertically developed **cumuliform** protuberances (some of which are taller than they are wide), which give the cloud a turreted appearance. See **altocumulus castellanus**.

catch—The amount of precipitation captured by a **rain gauge**.

catchment area—The area receiving waters feeding a **watercourse**. An area built specifically to collect rainfall. See **drainage area**.

Cb—Abbreviation for the cloud type **cumulonimbus**.

Cc—Abbreviation for the cloud type **cirrocumulus**.

CCL—Acronym for **convective condensation level**.

ceiling—The height above the earth's surface ascribed to the lowest **cloud layer** or **obscuring phenomena** when the **sky cover** is reported as **broken**, **overcast**, or **obscuration** and not classified "thin" or "partial." See **vertical visibility**.

ceilometer—An automatic, recording, cloud-height indicator. A light is projected upward onto the cloud base; the reflected light is detected by a photocell, and the height is determined by triangulation.

celestial equator—The projection of the plane of the geographical **equator** upon the **celestial sphere**.

celestial pole—One of the two points marking the intersection of the extension of the earth's spin axis with the **celestial sphere**. At present, the north celestial pole is located in the vicinity of Polaris (the "North Star").

celestial sphere—An apparent sphere of infinite radius with the earth at center. All celestial bodies (including the sun's path along the **ecliptic**) appear upon the *inner surface* of this sphere. Certain important features on the celestial sphere include the **celestial equator**, **celestial poles**, and **ecliptic**. Neglecting the effects of topography and atmospheric **refraction** near the **horizon**, for practical purposes half of this sphere (sometimes called *hemisphere* or *celestial dome*) may be considered visible from any point on the earth's surface at any time.

cellular cloud pattern—A **mesoscale** organization of **convection** in the form of a quasi-regular pattern of cloud cells. Such patterns may be composed of **open** or **closed cells** or both.

cellular convection—An organized, convective, **fluid** motion characterized by the presence of distinct **convection cells** or convective units, usually with upward motion (away from the heat source) in the central portions of the cell, and sinking or downward flow in the cell's outer regions. See **Benard cell**.

Celsius temperature scale—Used for most scientific purposes and having the numerical convenience of a 100° interval between the boiling point (100°C) and freezing point (0°C) of pure water at sea level air pressure. Named for Andres Celsius (1701–1744), a Swedish astronomer who devised this scale; occasionally still termed *centigrade scale*.

center of action—Any one of the large semipermanent **highs** and **lows** that appear on charts of mean sea level pressure. The main centers of action in the Northern Hemisphere are the **Icelandic low**, the **Aleutian low**, the **Azores high** and/or **Bermuda high**, the **Pacific high**, the **Siberian high** (in winter), and the **Asiatic low** (in summer). Since these systems dominate the atmospheric circulation, fluctuations in the nature of these centers are intimately associated with relatively widespread and long-term weather changes.

centimeter-gram-second system (**cgs system**)—A system of physical units based on the use of the centimeter, the gram, and the second as fundamental quantities of length, mass, and time, respectively. Compare with **(SI) International Units**.

central pressure—At any given instant, the **atmospheric pressure** at the center of a recognizable **high** or **low**; the highest pressure in a high, the lowest pressure in a low. Central pressure almost invariably refers to sea level pressure of systems on a **surface chart**.

central water—A near surface **water mass** characterized by warm temperatures and high **salinity**; found in subtropical and tropical oceans.

centrifugal force—A mythical force that is used to account for the tendency of an object that is following a curved path to be pulled radially outward. In reality, no such force exists because the object is exhibiting **inertial motion**; that is, the tendency of an object in straight-line unaccelerated motion is to continue that way. Curvature of the object is caused by the **centripetal force,** which is directed radially inward.

centripetal force—The agent producing **acceleration** of a particle moving in a curved path, directed toward the instantaneous center of curvature of the path. The magnitude of this force is directly proportional to the square of the instantaneous linear velocity and inversely proportional to the radius of curvature.

ceraunometer—Instrument used for counting the number of **lightning discharges** within a specific radius.

CFC—Abbreviation for **chlorofluorocarbons.**

cfs—Acronym for **cubic feet per second**.

cgs system—Acronym for **centimeter-gram-second system**.

chaff—Thin flat pieces of metallic foil ejected into the atmosphere to serve as a **target** for measuring the **winds aloft** by **radar**.

chain lightning (also known as ***bead lightning***)—A visible **lightning discharge** in a long zigzag or apparently broken line.

chance—In a probability of precipitation statement within a public forecast, a 30% to 50% chance of the occurrence of measurable precipitation, with scattered areal coverage.

change of phase—The transition of a substance from one state of matter (**solid**, **liquid**, or **gas**) to another, in which marked changes in physical properties and molecular structure occur; typically in meteorology, water substance is assumed to be involved. **Evaporation** and **condensation** are examples. **Latent heat** is involved with these phase changes.

change of state—Changes of the **variables of state**—that is, atmospheric pressure, density, temperature, and humidity in an atmospheric process—these properties being interrelated by the **ideal gas law** (**equation of state**) and by thermodynamic relations.

channel—(1) The deepest portion of a stream, bay, or strait. (2) Narrow range of wavelengths in the **electromagnetic spectrum** chosen to correspond to the response of a particular **radiometer**.

channel control—A condition whereby the **stage** of a stream is controlled only by **discharge** and the general configuration of the **stream channel** (i.e., the contours of its bed, banks, and flood plains).

channel storage—The water volume kept or stored within a specified portion of a **stream channel**.

chaos—A state of a system characterized by a controlled randomness produced by dynamic evolution that is highly dependent upon the **initial conditions**. Predictability is limited.

chaos theory—The general principle that internal instabilities cause complex behavior in a system. Because its **initial conditions** are imperfectly known, the earth–atmosphere system is said to be a chaotic dynamical system.

Charles' law—An empirical generalization that in a gaseous system at constant pressure, the temperature increase and the relative volume increase stand in approximately the same proportion for all so-called **perfect gases**. Named for Jacques Charles (1746–1823), a French chemist.

chemical energy—The form of **energy** produced or absorbed in the process of a chemical reaction. In such a reaction, energy losses or gains usually involve only the outermost electrons of the atoms or ions of the system undergoing change; here a chemical bond of some type is established or broken without disrupting the original atomic or ionic identities of the constituents. Chemical changes, according to the nature of the materials entering into the change, may be induced by heat (thermochemical), light (photochemical), and electric (electrochemical) energies.

chemosphere—The vaguely defined region of the upper atmosphere in which **photochemical reactions** take place. It is generally considered to include the **stratosphere** (or the top thereof) and the **mesosphere**, and sometimes the lower part of the **thermosphere**. This entire region is the location of a number of important photochemical reactions involving atomic oxygen O, molecular oxygen O_2, ozone O_3, hydroxyl OH, nitrogen N_2, sodium Na, and other constituents to a lesser degree.

chimney cloud—A **cumulus** cloud in the Tropics that has much greater vertical than horizontal extent. It frequently takes the form of a long ''neck'' protruding from the tops of a lower cloud mass where a locally strong convection current has penetrated the **temperature inversion**.

Chinook—The name given to the **foehn type** wind on the eastern slopes of the Rocky Mountains. The Chinook wind is warm and dry and generally blows from the southwest, but its direction may be modified by the pressure gradient and the topography. These winds can cause dramatic and rapid temperature rises (20°–40°F in 15 minutes) and the disappearance of wintertime **snow cover**; hence, this wind is often called a **snow eater**. **Adiabatic warming** of subsiding air and the advection of a warm Pacific air mass to replace the cold air contribute to this effect. See also **Santa Ana winds**.

Chinook arch—A characteristic cloud formation appearing as a bank of clouds over the Rocky Mountains, generally a flat layer of **altostratus**, heralding the approach of a **Chinook**. These **lenticular clouds** are found in the **lee wave** parallel to the mountain ridge.

chlorinity—The amount of chloride ion (and other halogen ions) in ocean water; expressed in parts per thousand by mass, such as grams per kilogram of **seawater**.

chlorofluorocarbons (**CFC**)—Synthetic compounds containing carbon and halogens including CFC-11, CFC-12, and CFC-13; used in the past as foam-blowing agents, aerosol propellants, refrigerants, and solvents because of assumed inertness. CFCs that enter the **stratosphere** are broken down by **ultraviolet** radiation releasing chlorine that reacts with and destroys **ozone**. Because of the CFC threat to the stratospheric ozone shield, manufacture and use of chlorofluorocarbons was banned in most countries by international agreement effective 1 January 1996.

chlorosity—The chloride content of one liter of **seawater**. It is equal to the **chlorinity** of the sample times its density at 20°C (68°F).

chromosphere (or *color sphere*)—A thin layer of relatively transparent **gases** (ionized hydrogen and helium) above the **photosphere** and below the **corona** of the sun. The chromosphere is best observed during a total solar eclipse when its emission spectrum may be studied.

Ci—Abbreviation for the cloud type cirrus.

circulation—The flow or motion of a **fluid** in or through a given area or volume.

circulation model—A simulation of atmospheric flow used to study its principal characteristics.

circumpolar vortex—Same as **polar vortex**.

circumpolar westerlies—The broad band of prevailing westerly winds in either hemisphere associated with the **circumpolar vortex**. See **westerlies** (2).

cirriform—Like *cirrus*; more generally, descriptive of **high clouds** composed of small particles, mostly ice crystals, which are fairly widely dispersed, usually resulting in relative transparency and whiteness and often producing **halo** phenomena not observed with other cloud forms. All species and varieties of **cirrus**, **cirrocumulus**, and **cirrostratus** clouds are cirriform in nature.

cirrocumulus (**Cc**)—A principal **high cloud** type (**cloud genus**) appearing as a thin, white patch, sheet, or layer of cloud without shading, composed of very small elements in the form of grains, ripples, etc., merged or separated, and more or less regularly arranged; most elements have an apparent width of less than one degree.

cirrostratus (**Cs**)—A principal **high cloud** type (**cloud genus**) appearing as a transparent whitish cloud veil, of fibrous or smooth appearance, totally or partially covering the sky, and generally producing **halo** phenomena. Cirrostratus may be produced by the merging of elements of cirrus; from **cirrocumulus**; from the thinning of **altostratus**; or from the **anvil cloud** of **cumulonimbus**.

cirrus (**Ci**)—A principal **high cloud** type (**cloud genus**) composed of ice crystals and detached cirriform elements in the form of white, delicate filaments, of white (or mostly white) patches, or of narrow bands. These clouds have a fibrous aspect and/or a silky sheen, or both. Cirrus often evolves from **virga** of **cirrocumulus** or **altocumulus**, or from the upper part of **cumulonimbus**. Cirrus may also result from the transformation of **cirrostratus** of uneven optical thickness, the thinner parts of which dissipate.

civil time—The time observed locally within a time zone, referenced to the civil day, reckoned from midnight to midnight. May include observance of daylight saving time. Differs from **solar time**.

civil twilight—The interval of incomplete darkness between **sunrise** (or **sunset**) and the time when the center of the sun's disc is 6° below the **horizon**; corresponds roughly to the minimum sky **illumination** required to carry on normal work out-of-doors without artificial illumination. Its actual duration varies considerably with latitude and time of year; in midlatitudes, its duration is on the order of 30 minutes. Compare with **nautical** and **astronomical twilight**.

Clausius–Clapeyron equation—The mathematical relationship between pressure and temperature in a system in which two phases of a substance are in **equilibrium**. Most often, this relationship is applied to relate the **saturation vapor pressure** of water substance and temperature. Named for Rudolph Clausius (1822–1888), a German physicist, and Benoit-Pierre-Emile Clapeyron (1799–1864), a French engineer.

clear—(1) In United States weather observing practice, the state of the sky when it is cloudless or when the **sky cover** is less than 0.1 (to the nearest tenth). (2) In United States climatological practice, the character of a day's weather when the average **cloudiness**, as determined from frequent observations, has been from 0.0 to 0.3 for the 24-hour period. Compare with **partly cloudy** and **cloudy**. (3) To change from a stormy or cloudy weather condition to one of no precipitation and decreased cloudiness. (4) In popular usage, the condition of the atmosphere when it is very transparent (as opposed to hazy, foggy, etc.) and accompanied by negligible cloudiness. In weather forecast terminology, the maximum cloudiness considered is about 0.2.

clear-air turbulence (CAT)—In aviation terminology, potentially dangerous chaotic airflow (**turbulence**) encountered by aircraft when flying through an air space devoid of clouds. **Thermals** and **wind shear** are the main causes of clear-air turbulence; a correlation also exists between the position of the **jet stream** and reported occurrences of high-altitude clear-air turbulence.

clear ice—Generally, a layer or mass of ice that is relatively transparent because of its homogeneous structure and small number and size of trapped air pockets.

climate—The total of all statistical weather information that helps to describe the variation of **weather** at a given place for a specified interval of time. In popular usage, the synthesis of weather; weather of some locality averaged over some time period (usually 30 years) plus statistics to include extremes in weather behavior recorded during that same period or for the entire period of record.

climate change—A significant change (i.e., a change having important economic, environmental, and social effect) in the climatic state of a locale or a large area; typically evident with a significant change in the mean values of a **weather element** (in particular, temperature or precipitation) in the course of a certain time interval, where the means are taken over periods on the order of a decade or longer.

climate model—Representation of the **climate system** based on the mathematical equations governing the behavior of the various components of the system and including treatments of key physical processes, interactions, and **feedback** phenomena.

climate snow line—The altitude above which a flat surface (fully exposed to sun, wind, and precipitation) would experience a net **accumulation** of snow over an extended period of time. Below this altitude, **ablation** would predominate. While this concept is largely theoretical in application, it corresponds closely to the actual

firn line of a **glacier** and to the average summer position of the **snow line** in mountainous terrain.

climate system—System consisting of the **atmosphere**, the **hydrosphere**, the **cryosphere**, the **lithosphere**, the **biosphere**, and the interactions between them, which under the effects of solar radiation received by the earth, determines the climate of the earth.

climate variability—(1) The inherent characteristic of **climate** that manifests itself in changes of climate with time. The degree of climate variability can be described by the differences between long-term statistics of **weather elements** calculated for different periods. In this sense, climate variability is the same as climate change. (2) Denotes deviations of climate statistics over a given period of time (e.g., a specific month, season, or year) from the long-term climate statistics relating to the corresponding calendar interval. In this sense, climate variability is measured by those deviations that are termed **anomalies**.

climatic classification—A systematic division of the earth's **climates** into worldwide groupings having common characteristics; climates are generally grouped according to the meteorological basis of climate, or the environmental effects of climate, such as vegetation distribution. The **Köppen's classification of climates** is a classic example.

climatic control—One of the relatively permanent factors that govern the general nature of the **climate** of a portion of the earth. They include (*a*) **solar radiation**, especially as it varies with latitude; (*b*) distribution of land-, and water masses, (*c*) elevation and large-scale topography; and (*d*) **ocean currents**. The **general circulation** or main wind systems sometimes are included, but they may better be considered a secondary control, since they themselves are controlled largely by the above factors.

climatic cycle—A long-period oscillation of **climate** that recurs with some regularity but is not strictly periodic.

climatic discontinuity—A **climate change** that consists of a rather abrupt and permanent change during the **period of record** from one average value to another.

climatic divide—A boundary between regions having different types of **climate**. The most effective climatic divides are the crests of mountain ranges. The boundary between two well-defined **ocean currents** may also act as a climatic divide.

climatic element—Any one of the properties or conditions of the atmosphere that together specify the physical state of **weather** or **climate** at a given place for any particular moment or period in time (e.g., temperature, humidity, precipitation).

climatic factor—See **climatic control**.

climatic fluctuation—A climatic inconstancy that consists of any form of systematic change, whether regular or irregular, except **climatic trend** and **climatic discontinuity**. It is characterized by at least two maxima (or minima) and one minimum (or maximum) including those at the end points of the record.

climatic forecast—A **forecast** of the future **climate** of a region—that is, a forecast of general weather conditions to be expected over a period of years.

Climatic Optimum (also called *Atlantic climate episode*)—An extended time interval between about 5000 and 7000 years ago when the global mean temperature was somewhat higher (perhaps 1°C greater) than at present.

climatic region—Region experiencing a fairly uniform **climate** according to specific criteria.

climatic trend—A **climate change** characterized by a reasonably smooth, monotonic increase or decrease of the average value of one or more **climatic elements** during the **period of record**.

climatic zone—A belt of the earth's surface within which the climate is generally homogeneous in some respect; an elemental region of a simple **climatic classification**, for example, a zone defined by the latitudinal distribution of the **climatic elements**. The expressions **polar**, **temperate**, **subtropical**, **tropical,** and **equatorial climate** are used to indicate the climatic zones that succeed each other from the pole to the equator.

climatization—Same as **acclimatization**. The process by which a living organism becomes adapted to a change in the environment.

climatography—A thorough, quantitative description of **climate**, particularly with reference to the tables and charts that show the characteristic values of **climatic elements** at a station or over an area.

climatological data—The many types of data—instrumental, historical (such as diaries or crop records), proxy (such as tree growth rings)—that constitute the major source of information for climate studies.

climatological forecast—A **weather forecast** for the next several days based upon the **climate** of a region instead of upon the dynamic implications of current weather. Considerations may be given to the climatic behavior of such synoptic weather features as **cyclones** and **anticyclones**, **fronts**, the **jet stream**, etc.

climatological standard normals—Averages of **climatological data** calculated for the following consecutive 30-year standardized intervals: 1 January 1901 to 31 December 1930; 1 January 1931 to 31 December 1960; 1 January 1961 to 31 December 1990.

climatology—The scientific study of **climate** and climate phenomena. In addition to the presentation of climatic data (**climatography**), it includes the analysis of the causes of differences of climate (**physical climatology**), and the application of climatic data to the solution of specific design or operational problems (**applied climatology**).

climograph (also known as *climagraph*)—A graph depicting climatic information of several **climatic elements** for a given locale, usually including the annual cycle of monthly mean temperatures and precipitation totals.

clinometer—An instrument for measuring angles of **inclination**; often used to measure cloud **ceiling** heights by triangulation.

closed cells—A **mesoscale** organization of **convective clouds**, which from a satellite perspective has the appearance of cloud patches of roughly equal size separated by cloud-free rings. Contrast with **open cells**.

closed circulation—The **circulation** of **fluid** within a closed **streamline**; a **vortex**.

closed drainage—In hydrology, an area in which the **surface flow** of water collects in "sinks" or lakes having no surface outlet.

closed high—An **anticyclone** that is completely encircled by at least one **isobar** or **contour**. (This means an isobar or contour line of any value, not necessarily restricted to those arbitrarily chosen for the analysis of the chart.) Also a **closed circulation** in the **anticyclonic** sense.

closed low—A **cyclone** that is completely encircled by at least one **isobar** or **contour**; a closed **cyclonic** circulation exists. Sometimes used to describe a **cutoff low**.

closed system—In thermodynamics, a system so chosen that no transfer of mass takes place across its boundaries; however, energy exchange may be permitted. For example, an **air parcel** undergoing a **saturation-adiabatic process**, as opposed to a **pseudoadiabatic expansion**. Compare with **isolated system**.

cloud—A visible aggregate of minute water droplets and/or ice crystals in the atmosphere above the earth's surface. A cloud differs from **fog** only in that the latter is, by definition, in contact with the earth's surface. Clouds form in the free atmosphere as a result of **condensation** or **deposition** of water vapor in ascending air.

cloud absorption—The attenuation of **electromagnetic radiation** through **absorption** by the water droplets, ice crystals, and water vapor within a cloud.

cloud amount—That fraction of the **sky cover** (reported in tenths of sky covered) that is attributed to a particular **cloud type**, or **cloud layer**; often used synonymously with **cloud cover**.

cloud bank—Generally, a fairly well-defined cloud mass observed at a distance; it covers an appreciable portion of the horizon sky but does not extend overhead.

cloud banner—Any long, narrow, unbroken line of cloud such as a **crest cloud** or an element of **billow cloud**.

cloud base—For a given **cloud** or **cloud layer**, the lowest level in the atmosphere at which the air contains a perceptible quantity of cloud particles (water droplets, ice crystals).

cloud burst—A popular term used to describe an excessive precipitation event, characterized by sudden, heavy, and often localized precipitation falling from a cloud; typically associated with a convective rain shower or thunderstorm.

cloud chart—A "Cloud Code Chart"; a cloud-observing guide showing photographs of typical cloud formations with corresponding official **cloud symbols**, codes, and names.

cloud classification—(1) A systematic scheme of distinguishing and grouping **clouds** according to their appearance, and, where possible, to their process of formation. The one in general use is based on a classification system introduced in 1803 by Luke Howard (1772–1864), a British pharmacist and meteorologist. This system was adopted by the World Meteorological Organization and published in the *International Cloud Atlas* (1956). This classification is based on the determination of (*a*) *genera*, the main characteristic forms of clouds; (*b*) *species*, the peculiarities in shape and differences in internal structure of clouds; (*c*) *varieties*, special characteristics of arrangement and transparency of clouds; (*d*) *supplementary features and accessory clouds*, appended and associated minor cloud forms; and (*c*) *mother-clouds*, the origin of clouds if formed from other clouds. The 10 cloud genera are **cirrus**, **cirrostratus**, **cirrocumulus**, **altocumulus**, **altostratus**, **nimbostratus**, **stratocumulus**, **stratus**, **cumulus**, and **cumulonimbus**. (2) A scheme of classifying clouds according to their usual altitudes or vertical extent. Three **cloud levels** (or *étages* of the **troposphere**) are distinguished: **high**, **middle**, and **low**. High clouds include cirrus, cirrocumulus, cirrostratus, occasionally altostratus, and the tops of cumulonimbus. The *middle clouds* are altocumulus, altostratus, nimbostratus, and portions of cumulus and cumulonimbus. The *low clouds* are stratocumulus, stratus, most cumulus and cumulonimbus bases, and sometimes nimbostratus. An additional class within this designation scheme that is often used is the **vertically developed clouds**, extending through the three étages, and including cumulus and cumulonimbus. (3) A scheme of classifying clouds according to their particulate composition; namely, **water clouds**, **ice crystal clouds**, and **mixed clouds**. The first are composed entirely of water droplets (ordinary and/or supercooled), the second entirely of ice crystals, and the third a combination of the first two. Of the cloud genera, only cirrostratus and cirrus are almost always ice crystal clouds; cirrocumulus can also be mixed; and only cumulonimbus are always mixed. Altostratus nearly always are mixed, but occasionally can be ice crystal. All the rest of the genera are usually water clouds, occasionally mixed; altocumulus, cumulus, nimbostratus, and stratocumulus.

cloud condensation nuclei—Small **aerosols** that serve as the sites upon which water vapor condenses in the atmosphere; typically, these nuclei are **hygroscopic**. Contrast with **freezing nuclei**.

cloud cover—That portion of the **sky cover** that is attributed to clouds, usually measured in tenths of sky covered. Often used synonymously with **cloud amount**.

cloud discharge—A **lightning discharge** occurring between a positive charge center and a negative charge center, both of which lie in the same cloud. Often reported as "*in cloud*" or "*intracloud*" lightning.

cloud droplet—A particle of liquid water from approximately 4 to 100 micrometers in diameter, formed by condensation of atmospheric water vapor, and suspended in the atmosphere with other droplets to form a cloud. Smaller than **drizzle** or **raindrops**.

cloud element—A distinguishable entity of a **cloud**, usually used for classification or tracking.

cloud formation—(1) The process by which various types of **clouds** are formed. In most cases, cloud formation involves **adiabatic cooling** of ascending **humid air**. In rare exceptions, such as in the case of a **banner cloud** or **fog** (which may produce **stratus**), the cooling may occur as a result of other processes. (2) A particular arrangement of clouds in the sky, or a striking development of a particular cloud.

cloud genus (pl. *genera*)(also known as *cloud type*)—The main characteristic form of a cloud used in its identification. See **cloud classification** (1).

cloud height—(1) In weather observations, the **altitude** of the **cloud base** above local terrain. See **ceiling**. (2) The **altitude** of the cloud top above local terrain or above **mean sea level**. (3) The vertical distance from the cloud base to the cloud top; more commonly referred to as the "*thickness*" or "*depth*" of the cloud.

cloud layer—An array of **clouds**, not necessarily all of the same type, whose **cloud bases** appear to be at approximately the same altitude or level. It may be either continuous or composed of detached **cloud elements**.

cloud level—(1) A layer in the atmosphere in which are found certain **cloud genera**. Three such levels are usually defined: **high**, **middle**, and **low**. (2) At a particular time, the layer in the atmosphere bounded by the limits of the bases and tops of an existing cloud form.

cloud modification—Any process by which the natural course of development of a **cloud** is altered by artificial means. The goal of cloud modification may be either the dissipation of the cloud or the stimulation of precipitation. See **cloud seeding**.

cloud movement—The lateral displacement of a **cloud element**; an observation of recognizable cloud elements over time from ground or satellites used to deduce **winds aloft**, based on the assumption that a **cloud element** can serve as a tracer for ambient wind flow at that **cloud level**.

cloud particle—An airborne particle of water substance, either a **cloud droplet** of liquid water or an **ice crystal**; one of the many composing a **cloud**.

cloud physics—The body of knowledge concerned with physical properties of **clouds** in the atmosphere and the processes occurring therein.

cloud seeding—Any technique carried out with the intent of adding to a cloud certain particles that will alter the natural development of that cloud, generally leading to precipitation. See **cloud modification**.

cloud sheet—Particular arrangement of **clouds** forming a continuous and relatively thin layer of great horizontal extent.

cloud shield—In **synoptic meteorology**, the principal and extensive cloud layer often found on the cold air side of a typical **wave cyclone** and frontal system. The maximum areal coverage is usually found over the region in advance of the **warm front**, and the minimum behind the **cold front**. Within the area of the cloud shield, a smaller *precipitation shield* is usually found.

cloud species—A subdivision of a **cloud genus** based upon the cloud shape or internal structure; examples include **cumulus humulus** or **cumulus congestus**.

cloud street—A group of **cloud elements**, especially from **cumuliform** cloud types, arranged in lines or rows roughly parallel to the **wind direction**, especially from a satellite perspective. To a ground-based observer, these rows may appear, on the account of perspective, to converge toward a point. When the bands pass completely across the sky, the cloud street may appear to converge to two opposite points on the horizon.

cloud symbol—One of a set of specified standard graphical images (ideograms) that represent the various cloud types of greatest significance or those most commonly observed. Cloud symbols are entered on a **surface weather map** as part of a **station model**.

cloud type—See **cloud genus**.

cloudburst—In popular terminology, any sudden and heavy fall of rain, almost always of the **shower** type. An unofficial criterion sometimes used specifies a rate of fall equal to or greater than 100 mm (3.94 inches) per hour.

cloudiness—Same as **cloud cover** but usually used in a very general sense.

cloud-to-cloud discharge—A **lightning discharge** occurring between a negative electrostatic charge center of one cloud and a positive charge center of a second cloud.

cloud-to-ground discharge—A **lightning discharge** occurring between an electrostatic charge center (usually negative) in the cloud and a center of opposite charge at the ground.

cloudy—(1) In popular usage, the state of the weather when clouds predominate at the expense of sunlight, or obscure the stars at night. In weather forecast terminology, expected **cloud cover** of about 0.7 or more warrants the use of this term; contrast with **clear** and **partly cloudy**. (2) In United States climatological practice, the character of a day's weather when the average **cloudiness**, as determined from frequent observations, has been 0.8 or more for the 24-hour period.

clutter—A **radar** term for the undesirable echoes observed on a **radar screen**. The usual reference here is to **ground clutter** from permanent ground-based **targets** near the radar.

coagulation—(1) Same as **accretion**. (2) Less frequently, in cloud physics, any process that converts a cloud's numerous small **cloud droplets** into a smaller number of larger precipitation particles. When so used, the term is employed in analogy to the coagulation of any **colloidal system**.

coalescence—In cloud physics, the merging of two water droplets into a single larger drop; a feature of the **collision-coalescence process**.

coalescence efficiency—The fraction of all collisions between water droplets of a specified size that results in actual merging of the two droplets into a single larger drop.

coalescence process—See **collision-coalescence process**.

coastal current (also known as ***nearshore*** or ***offshore current***)—A flow of water (**drift** or current) paralleling the shore, seaward of the **surf zone**. These currents are usually caused by tides, winds, or redistribution of water mass in coastal regions, especially from river discharge.

coastal flooding—The inundation of land areas along a coast by seawater forced above the normal tidal conditions by an event such as a storm surge or tsunami.

coastal flood warning—Issued by the National Weather Service to warn residents of coastal areas that land areas along the coast will be inundated by seawater above the typical tide action.

coastal flood watch—Issued by the National Weather Service to alert coastal residents of the possibility of the inundation of land areas along the coast within the next 12–36 hours.

col (also known as a *saddle point*)—In meteorology, the point of intersection of a **trough** and a **ridge** in the pressure pattern of a **weather map**. It is the point of relatively lowest pressure between the two **highs** and the point of relatively highest pressure between two **lows**.

cold air advection (or *cold advection*)—The transport of cold air from a region of low temperatures to a region of high (warm) temperatures by wind motion. Contrast warm air advection.

cold air damming—The shallow layer of cold air trapped between the Atlantic coast and the Appalachian Mountains by anticyclonic circulations.

cold air funnels—Conical-shaped cloud protuberances from **cumuliform clouds** that remain aloft (**funnel clouds**) in unstable cold air masses, not usually associated with **thunderstorm** activity.

cold air outbreak—Same as **polar outbreak**.

cold desert—Same as **arctic desert**.

cold dome—A cold convex-shaped **air mass**, considered as a three-dimensional entity and typically bounded by a well-defined **frontal zone**.

cold front—Any nonoccluded **front**, or portion thereof, that moves so that the colder air replaces the warmer air; that is, the "leading edge" of a relatively cold **air mass**. Although some **occluded fronts** exhibit this characteristic, they are more properly termed **cold occlusions**.

cold high—At a given level in the atmosphere, any **anticyclone** that is generally characterized by colder air near its center than around its periphery. Opposite of a **warm high**.

cold low—At a given level in the atmosphere, any **cyclone** that is generally characterized by colder air near its center than around its periphery. Opposite of a **warm low**. A significant case of a cold low is that of a **cutoff low**, characterized by a completely isolated pool of cold air within its **vortex**.

cold occlusion (or *cold occluded front*)—A type of occlusion with an **occluded front** where the cold air behind the front is colder than the cool air ahead of it. The original **cold front** undercuts the **warm front**, with the coldest air replacing the less cold air at the earth's surface. This type of occlusion is more common than the **warm occlusion**.

cold pool—A region, or "pool," of relatively cold air surrounded by warmer air. Opposite of a **warm pool**. Any large-scale mass of cold air; a cold air mass or **cold dome**.

cold tongue—In **synoptic meteorology**, a pronounced equatorward extension or protrusion of cold air.

cold wall—The steep water–**temperature gradient** between the **Gulf Stream** and (*a*) the **slope water** inshore of the Gulf Stream or (*b*) the **Labrador Current**.

cold wave—(1) A rapid fall in temperature within 24 hours to temperatures requiring substantially increased protection to agriculture, industry, commerce, and social activities. Therefore, the criterion for a cold wave is twofold: the rate of temperature fall and the minimum to which it falls. The latter depends upon region and time of year. (2) Popularly, an interval of very cold weather.

cold-core anticyclone—Same as **cold high**.

cold-core cyclone—Same as **cold low**.

cold-core high—Same as **cold high**.

cold-core low—Same as **cold low**.

collection efficiency—In general, the fraction of all particles initially moving on a collision course with a given impactor that actually do collide with and remain adhered to that impactor.

collision efficiency—The fraction of all water drops that, initially moving on a collision course with respect to other drops, do actually collide (make surface contact) with the other drops.

collision-coalescence process—The growth of **raindrops** by the collision and **coalescence** of **cloud droplets**.

colloidal system—An intimate mixture of two substances one of which, called the *dispersed phase* (or colloid), is uniformly distributed in a finely divided state through the second substance, called the *dispersion medium* (or *dispersing medium*). The dispersion medium may be a **gas**, a **liquid**, or a **solid**; the dispersed phase may also be any of these, with the exception that one does not speak of a colloidal system of one gas in another.

color temperature—An estimate of the temperature of an incandescent body, determined by observing the wavelength at which it is emitting with peak intensity (its color) and using that wavelength in **Wien's displacement law**; the color temperature of a **blackbody** radiator is the same as the actual radiative temperature.

Colorado low—A **cyclone** that makes its first appearance as a definite center in the vicinity of Colorado on the eastern slopes of the Rocky Mountains. It is, in most aspects, analogous to the **Alberta low**.

columnar ice crystal—A relatively short prismatic **ice crystal**, either solid or hollow. Its end may be plane, pyramidal, truncated, or hollow. Pyramids and combinations of columns are included in this class.

comber—A large ocean wave with a high, breaking crest.

comfort curve—A line drawn on a graph of air temperature versus some function of humidity (usually **wet-bulb temperature** or **relative humidity**) to show the varying conditions under which the average sedentary person feels the same degree of comfort; a curve of constant comfort.

comfort index—See **temperature–humidity index**.

comfort zone—The ranges of indoor temperature, humidity, and air movement, under which most persons enjoy mental and physical well-being. In the United States the comfort zone with normal ventilation lies between air temperatures of about 68°F (20°C) in winter to 79°F (26°C) in summer at a **relative humidity** of 60%, and from 69°F (20°C) in winter to 81°F (27°C) in summer at a dewpoint of 36°F (2°C), giving an **effective temperature** within a few degrees of 72°F (22°C) in winter and 77°F (25°C) in summer.

comma cloud—The widespread comma-shaped cloud system typically associated with a mature **midlatitude cyclone** as viewed from a satellite image.

complex low—An area of low atmospheric pressure within which more than one low pressure center is found.

composite flash—A **lightning discharge** consisting of a series of distinct **lightning strokes** all of which follow the same or nearly the same channel. Successive strokes follow each other at intervals of about 0.05 seconds.

compressible fluid—A **fluid** whose density changes with a change in pressure. The atmosphere is considered to be compressible since both density and pressure change in the vertical. Contrast with **incompressible fluid**.

computational instability—A phenomenon that can arise in **numerical weather prediction** where errors in approximate methods rapidly increase. It results from interactions of grid size and time steps employed in the calculations.

concentration time—The time required for water from storm **runoff** to travel from the most remote portion of a river basin to the basin outlet. It is not a constant, but varies with quantity of flow and channel conditions.

condensation—The physical process by which a **vapor** becomes a **liquid**. Opposite of **evaporation**. Any process in which a solid forms directly from its vapor is termed **deposition**. **Latent heat of condensation** is released from the water in the process, typically warming the environment.

condensation level pressure—The pressure at which a humid **air parcel** expanded dry adiabatically reaches saturation; the pressure at the **Lifted Condensation Level (LCL)**.

condensation nucleus—A particle, either **liquid** or **solid**, upon which condensation of water vapor begins in the atmosphere; also may refer to **cloud condensation nucleus**. See **nucleation**.

condensation temperature—The temperature at which a humid **air parcel** expanded dry adiabatically reaches saturation; occurs at the **Lifted Condensation Level (LCL)**.

condensation trail (or *contrail*)—A cloud-like streamer frequently observed to form behind aircraft flying in clear, cold air.

conditional instability—The state of a column of air in the atmosphere when its **lapse rate** of temperature is less than the **dry-adiabatic lapse rate** (9.8°C per kilometer or 5.4°F per 1000 feet) but greater than the **saturation-adiabatic lapse rate** (4°–7°C per kilometer or 2°–3.5°F per 1000 feet). With respect to the vertical displacement of an **air parcel** (see **parcel method**), the saturated parcel will continue away, such that the layer is considered unstable to saturated adiabatic processes, while the unsaturated parcel will return, such that the layer is considered stable to unsaturated (dry) adiabatic processes. Contrast with **absolute instability**.

conduction—The transfer of **energy** (e.g., heat) within and through a medium (*conductor*) from high (hot) to low (cold) energy regions by means of internal particle or molecular activity, and without any net external motion of the medium. Heat and electricity conduction are examples. Conduction is to be distinguished from **convection** (of heat) and **radiation** (of all electromagnetic energy). See **heat transfer**.

conduction current—The migration of electrically charged particles in a gaseous medium acted upon by an external **electric field**.

conductivity—A unit measure of electrical conduction; the facility with which a substance conducts electricity. See **heat conductivity**.

cone of depression—A depression, roughly conical in shape and convex upward, formed in a **water table** or **piezometric surface**, by the withdrawal of water from a well; defines the area of influence of the well.

confluence—The rate at which adjacent flow is converging along an axis oriented perpendicular to the flow at the point in question. Opposite of **diffluence**. Streamlines become closer in direction of flow. This confluence condition may cause **convergence**.

coning—A pollution **plume** that spreads outward from the downwind axis in a conical shape because of weakly stable atmospheric conditions extending upward through the entire layer. Contrast with **fanning** and **trapping**.

conservation of angular momentum—The principle that the product of an object's mass, speed, and radial distance of rotation is constant in the absence of rotational forces applied to the system. Thus, as the radial distance of rotation decreases, the speed of rotation increases and vice versa.

conservation of energy—The principle that the total energy of an **isolated system** remains constant. That is, energy cannot be created or destroyed but can be converted from one form to another.

conservation of mass—The principle that states that mass cannot be created or destroyed but only transferred from one volume to another. See **equation of continuity**.

conservation of momentum—The principle that, in the absence of forces, absolute linear **momentum** is a property that cannot be created or destroyed. The product of an object's linear velocity and mass remains constant.

conservation of vorticity—The statement that in the horizontal flow of a **fluid**, the vertical component of **absolute vorticity** of each individual fluid particle remains constant.

conservative property—An attribute of an object or system whose values do not change in the course of a particular series of events. Properties can be judged conservative only when the events (processes) are specified; also, properties that are conservative for a whole system may or may not be conservative for its parts and conversely. Meteorologists and oceanographers often use certain conservative properties of an **air mass** or **water mass** as tracers.

constant-height chart—A **synoptic chart** for any surface of constant geometric altitude above mean sea level (a **constant-height surface**), usually containing plotted data and analyses of the distribution of such variable **weather elements** as pressure, wind, temperature, and humidity at that altitude. Currently, the only commonly analyzed constant-height chart is the **surface chart** (or *sea level chart*). Essentially all of the operational **upper-air charts** are **constant-pressure charts**.

constant-height surface—In meteorology, a surface of constant geometric or **geopotential altitude** measured with respect to **mean sea level**.

constant-level balloon—A balloon designed to float at a **constant-pressure level**.

constant-pressure chart (also known as *isobaric chart*)—The **synoptic chart** for any **constant-pressure surface**, usually containing plotted data and analyses of the distribution of such variable **weather elements** as height (specifically, **geopotential height**) of the surface, wind, temperature, and humidity. Constant-pressure charts are most commonly known by their pressure value; for example, the 1000 hectopascal (which closely corresponds to the **surface chart**), the 850 hectopascal, 700 hectopascal, and 500 hectopascal charts, etc.

constant-pressure surface (also known as *isobaric surface*)—In meteorology, a surface along which the **atmospheric pressure** is everywhere equal at a given instant.

consumptive use—(1) Traditionally, the total amount of water taken up by vegetation for **transpiration** or building of plant tissue, plus the unavoidable **evaporation** of **soil moisture**, snow, and intercepted precipitation associated with the vegetal growth. (2) Use of water for industrial or community consumption.

contaminant—A substance in the air that is not a natural constituent of the atmosphere, and that may be harmful to humans, animals, plants, and structures. Commonly called **air pollutants**.

continent—One of the seven large continuous areas of planet Earth that rise above the deep sea ocean floor to form the land surface; this area includes the shallow submerged **continental margins**.

continental air—A type of **air mass** whose characteristics are developed over a large land area and which, therefore, has the basic continental characteristic of relatively low water vapor content. Contrast with **maritime air**.

continental borderland—A submarine plateau or irregular area adjacent to a continent, with depths greatly exceeding those on the **continental shelf**, but not as great as in the deep oceans.

continental climate—The **climate** that is characteristic of the interior of a landmass of continental size. It is marked by large annual, daily, and day-to-day ranges of temperature, low relative humidity, and (generally) by a moderate or small and irregular rainfall. The annual extremes of temperature typically occur within a month after the **solstices**. In its extreme form, a continental climate gives rise to **deserts**. Contrast with **maritime climate**.

continental divide—An imaginary line separating adjacent river systems that flow into different oceans. In North America, the name is usually associated with the mountain ridges in the Rocky Mountains separating the water from the Pacific and the Atlantic and the Gulf of Mexico and Hudson Bay.

continental drift—The extremely slow lateral motion of landmasses across the face of the globe; continents are components of huge tectonic plates. Originally proposed in 1912 by Alfred L. Wegner (1880–1930), the German polar meteorologist. See **plate tectonics**.

continental glacier—A continuous sheet of **land ice** that covers a very large area and moves outward in many directions. This type of ice mass is so thick as to mask the land surface contours, in contrast to the smaller and thinner **highland ice**. A characteristic of the **ice ages**; the antarctic **ice sheet** is a present-day example.

continental high—A general area of high atmospheric pressure that, on charts of mean sea level pressure, overlies a continent during the winter. The only really pronounced example is the **Siberian high**.

continental margin—A submerged zone separating the emerging continents from the surrounding deep ocean floor, to include the **continental shelf** and **continental slope**.

continental platform—The offshore zone that includes both the **continental shelf** or **continental borderland** and the **continental slope**.

continental shelf—The submerged zone around the continents extending from the low-water mark seaward to where a marked increase in slope to greater depths indicates the **continental slope**.

continental slope—The submerged declivity from the outer edge of the **continental shelf** or **continental borderland** into greater depths of the ocean.

continentality—The degree to which the climate of a point on the earth's surface is in all respects subject to the influence of a landmass; in climatology, a numeric index based upon the annual temperature range. The reverse of **oceanicity**.

continuity—The property of a **field** (e.g., temperature, pressure) such that neighboring values of a parameter differ only by an arbitrarily small amount if they are sufficiently close in space and/or time. Contrast with **discontinuity**.

continuity equation—See **equation of continuity**.

contour—An **isopleth** or line, typically drawn on a map or chart, joining points of equal value (e.g., altitude, temperature) on a particular surface; usually refers to the *height contours* connecting equal altitudes (specifically, **geopotential heights**) of a three-dimensional **isobaric** surface above the mean sea level reference level.

contour map—A map that shows the configuration of a three-dimensional surface by means of **contour** lines drawn at regular intervals; in meteorology, a map with **geo-potential height** contours of a constant pressure surface are usually assumed.

contrail—Contraction of **condensation trail**.

convection—In general, mass motions within a **fluid** resulting in **transport** and **mixing** of the properties of that fluid. Convection, along with **conduction** and **radiation**, is a principal means of **energy transfer**. In meteorology, generally applied to atmospheric motions that are predominantly vertical, distinguished from **advection**. The term may refer to upward-motions. See also discussion in **heat transfer**.

convection cell—A systematic circulation pattern in a body of air or water arising from density differences within the **fluid**.

convection current—Any current of air involved in **convection**. In meteorology, this is usually applied to the upward-moving portion of a convection circulation, such as a **thermal** or the **updraft** in **cumuliform** clouds.

convective activity—General term for manifestations of **convection** in the atmosphere, alluding particularly to the development of **convective clouds** and resulting weather phenomena, such as **showers**, **thunderstorms**, **squalls**, **hail**, **tornadoes**, etc.

convective cloud—A **cumuliform** cloud that owes its vertical development, and possibly its origin, to **convection**. **Cumulus** and **cumulonimbus** are examples.

convective condensation level (**CCL**)—The level in the atmosphere to which an **air parcel**, if heated sufficiently from below, will rise dry adiabatically, without becoming colder than its environment until the parcel just becomes saturated; this level may correspond to the altitude of the bases of **cumuliform clouds**. To attain the CCL, the surface air must be heated to the convective temperature. Contrast with **Lifted Condensation Level** (**LCL**).

convective precipitation—Precipitation from **convective clouds**; generally considered to be synonymous with **showers**.

convective region—Generally, an area that is particularly favorable for the formation of **convection** in the lower atmosphere; or one characterized by convective activity at a given time.

convective temperature—The lowest temperature to which the surface air must be heated before a parcel can rise dry adiabatically to its **lifting condensation level** without ever being colder than the environment. This temperature is a useful parameter in forecasting the onset of convection.

convective theory of cyclogenesis—A theory of **depression** formation proposing that the upward convection of air due to surface heating can be of sufficient magnitude and duration to cause the inflowing air near the earth's surface to acquire appreciable **cyclonic** rotation. This theory was developed during the nineteenth century. Compare with **wave theory of cyclones**.

convergence—A **fluid** flow pattern that brings about a net inflow of fluid elements into a region. Opposite of **divergence**, or mathematically, the negative of divergence.

This convergence must be accompanied by compensating vertical motion. See **confluence**.

convergence line—Line along which maximum horizontal **convergence** of the airflow is occurring. If mass convergence takes place in a plane near the earth's surface, the incoming air must rise at the convergence line. Hence, the lines are often associated with clouds.

convergence zone—A near-surface region of the ocean where **water masses** having different properties meet or converge; sinking of surface water usually results. Examples include the **Arctic**, **Antarctic**, and **Subtropical Convergence**.

cooling degree-day unit—An index of the energy requirements for air conditioning or refrigeration; one cooling degree-day unit is given for each degree that the **mean daily temperature** departs above the base of 65°F. Contrast with **heating degree-day unit**.

cooperative observer—An unpaid observer who maintains a climatological station for the National Weather Service. The usual instruments furnished such an observer are **maximum** and **minimum thermometers** and a **nonrecording precipitation gauge**. The data obtained from these stations are included in the **climatography** of the region.

coordinate system—A frame of reference that is used to specify the location or motion of a particle in space; the familiar north–south, east–west, up–down frame of reference is an example.

core sample—A sample of rock, soil, snow, or ice obtained by driving a hollow tube into the medium and withdrawing it with its contained sample or "core." In general, the aim of core sampling is to obtain a specimen in its undisturbed natural state for subsequent analysis.

Coriolis effect—An apparent force, relative to the earth's surface, that causes deflection of moving objects to the right in the Northern Hemisphere and to the left in the Southern Hemisphere due to the earth's rotation. Atmospheric and oceanic motions are influenced by the Coriolis effect. Named for Gustav Gaspard de Coriolis (1792–1843), a French mathematician who published a quantitative mathematical work on the subject in 1835.

Coriolis parameter—Twice the component of the earth's **angular velocity** about the local vertical. Because the earth is in rigid rotation, the Coriolis parameter is equal to the component of the earth's **vorticity** about the local vertical; varies as the sine of the **latitude**.

corona—(1) A **photometeor**, identified as a set of one or more prismatically colored rings of small radii, concentrically surrounding the disk of the sun, moon, or other luminary when veiled by a thin cloud. The corona is due to **diffraction** of light by numerous uniformly sized water droplets, especially in thin low- and midlevel **stratiform** clouds. It can be distinguished from the relatively common **halo of 22 degrees** by the much smaller angular diameter of the corona, which is often only a few degrees, and by its color sequence, which is from blue inside to red outside, the reverse of that in the 22° halo. (2) The pearly outer envelope of the sun consisting of hot (1–4 million °C) rarefied gases, extending millions of kilometers out from the sun. This layer is observed during a solar eclipse.

corona discharge—A luminous, and often audible, electric discharge that is intermediate in nature between a spark discharge (with, usually, its single discharge channel) and **a point discharge** (with its diffuse, quiescent, and nonluminous character). It occurs from objects, especially elevated pointed ones, when the **electric field strength** near their surfaces attains a value near 1000 volts per centimeter. See **St. Elmo's fire**.

corrected altitude—The **indicated altitude** corrected for temperature deviation of the air column from the **standard atmosphere**; an approximation of **true altitude**.

cosmic rays—The aggregate of extremely high energy subatomic particles (e.g., protons) that bombard the earth's atmosphere from outer space; their **absorption** takes place in the lowest 20 kilometers of the atmosphere.

cotidal hour—The average interval of time between the moon's passage over the meridian of Greenwich and the following **high water** at a specified place.

countercurrent—An ocean current flowing adjacent to another current but in the opposite direction.

counterglow—Faint spot of light, round or elongated, in the night sky at the **antisolar point**, linking the east and west **zodiacal lights**.

course—The direction of a line over the earth with reference to **true** or **magnetic north**. Compare with **heading**.

crepuscular rays—Literally, "twilight rays"; alternating lighter and darker bands (rays and shadows) that appear to diverge in fanlike array from the sun's position at about **twilight**; a **twilight phenomenon**. The effect can also be produced by sunlight passing between vertical **cloud elements** in a hazy atmosphere during daylight.

crest (or ***wave crest***)—Highest part of a **wave**. See also **ridge**. Opposite of **trough** (1).

crest cloud—A type of **standing cloud** that forms along a mountain range, either on the ridge or slightly above and to **leeward** of it, and remains in the same position relative to the ridge. Its process of formation and maintenance is identical to that of the **cap cloud**.

crest stage—The highest water level (**stage**) reached at a point along a stream culminating a rise by waters of that stream.

criteria air pollutants—Contaminants in the ambient air for which the U.S. Environmental Protection Agency (EPA) has set maximum acceptable concentration based primarily on the perceived threat to human health. See **National Ambient Air Quality Standards**.

critical depth—In a specified stream **channel**, the water depth at which the **specific energy** is the minimum for a given flow rate, or the **discharge** is a maximum for a given specific energy.

critical depth control—A condition in a stream where, at a certain point, the water depth passes from above **critical depth** to below critical depth.

critical flow—The flow condition of a **fluid** system when one of the fundamental nondimensional parameters has a critical value.

critical point—The thermodynamic state in which **liquid** and **gas** phases of a substance coexist in **equilibrium** at the highest possible temperature. At higher temperatures than the critical, no liquid phase can exist. For water, the critical point is a saturation vapor pressure of 221 000 hectopascals (millibars), a temperature of 647 K, and a specific volume of 3.10 grams per cubic centimeter.

Cromwell Current—A name for the **Equatorial Undercurrent;** named for Townsend Cromwell, an oceanographer who studied this current.

cross section—A graphical two-dimensional representation of a three-dimensional entity. In **weather analysis** and **forecasting**, a vertically oriented surface in the atmosphere portraying the **weather elements** observed simultaneously at various altitudes along a given horizontal path; this surface extends from the earth's surface to a given level, such as the stratosphere, and is specified either in vertical height or pressure coordinates.

crosswind—A wind having a component that is directed perpendicularly to the **course** (or **heading**) of an exposed, moving object; more popularly, a wind that predominantly acts in this manner.

cryosphere—That part of the planet earth covered with permanent ice; it interacts with the other parts of the planet: the **atmosphere**, **hydrosphere**, **lithosphere**, and **biosphere**.

crystal—A **solid** of known or determinable, ideally homogeneous composition. In physics, the term crystal is synonymous with solid or true solid—that is, all solid matter with a regular internal structure (crystal lattice) exhibiting homogeneity and symmetry in its structure and chemical composition.

crystallization—The process of solidification of materials that have homogeneous molecular orientation (as opposed to amorphous substances); the process of **crystal** formation.

Cs—Abbreviation for the cloud type cirrostratus.

Cu—Abbreviation for the cloud type cumulus.

cubic foot per second (cfs)—A unit of discharge commonly used in hydrology and hydraulics, representing a volume of one cubic foot passing a given point during 1 second. This rate is equivalent to approximately 7.48 gallons per second.

cumuliform—Like *cumulus*; generally descriptive of all clouds, typically convective in origin, with the principal characteristic of vertical development in the form of rising mounds, domes, or towers. This is the contrasting form to the horizontally extended **stratiform** types.

cumulonimbus (Cb)—A principal **cloud type** (**cloud genus**), exceptionally dense and vertically developed, occurring either as isolated clouds or as a line or wall of clouds with separated upper portions; also called a **vertically developed cloud**. These clouds appear as mountains or huge towers, at least a part of the upper portions of which are usually smooth, fibrous, or striated, and almost flattened. This part often spreads out in the form of an **anvil** or vast plume, and is popularly known as a **thunderhead**. Under the base of cumulonimbus, which often is very dark, **virga**, precipitation and low, ragged clouds, either merged with it or not, frequently exist.

cumulus (**Cu**)—A principal **cloud type** (**cloud genus**) in the form of individual, detached vertically developed **cloud elements** that are generally dense and possess sharp nonfibrous outlines. These elements develop vertically, appearing as rising mounds, domes, or towers, the upper parts of which often resemble a cauliflower. The sunlit parts of these clouds are mostly brilliant white; their bases are relatively dark and nearly horizontal. **Cumulus congestus** and **cumulus humilis** are **cloud species** within the genus.

cumulus congestus—A strongly sprouting **cumulus species** with generally sharp outlines and, sometimes, with a great vertical development, it is characterized by its cauliflower or tower aspect, of large size. These clouds are often reported as **towering cumulus**. Mainly in the Tropics, cumulus congestus may produce abundant precipitation. It may also occur in the form of very high towers, the tops of which are formed by cloudy puffs that, detaching themselves successively from the main portion of the cloud, are carried away by the wind, then disappear more or less rapidly, sometimes producing **virga**. The tops do not have the filmy (glaciated) **anvil** that is characteristic of a **cumulonimbus** cloud.

cumulus humilis—A **species** of **cumulus** characterized by a small vertical development and by a generally flattened appearance. Its vertical growth is usually restricted by the existence of a **temperature inversion** in the atmosphere; this, in turn, explains the unusually uniform height of the cloud tops of this cumulus species.

cup anemometer—A **rotation anemometer** whose axis of rotation is vertical. Cup anemometers usually consist of three or four hemispherical or conical cups mounted with their diametrical planes vertical and distributed symmetrically about the axis of rotation. The rate of rotation of the cups, which is a measure of the **wind speed**, is determined indirectly by gearing a mechanical or electrical counter to the shaft.

curie—That amount of any radioactivity that yields 3.700×10^{10} disintegrations per second; also the amount of radioactive material having one curie of radioactivity. Named for M. Curie.

current chart—A map of a water area depicting ocean current data by **current roses**, vectors, or other means.

current ellipse—A two-dimensional display of the change in the speed and direction of a **rotary current** with respect to time over a **tidal cycle**, typically a 24-hour **tidal day**. Hourly current velocity readings are plotted by radius vectors from the central origin of the graph. Line segments joining the end points of these vectors can be drawn to depict the turning of the rotary current with time, usually in a clockwise direction in the Northern Hemisphere, which over a tidal cycle would form a complete ellipse.

current meter—An instrument used for measuring and indicating the speed and/or direction of flowing water (**current**); most current meters utilize rotary cups or propellers.

current rose—A diagram that indicates, for a given ocean area, the average percentage of current **set** toward each of the principal compass points utilizing arrows to indicate current direction. The frequency distribution of **drifts** is sometimes also indicated, thereby showing prevailing current data.

current tables—Annual tables of daily predictions of times and velocities of **maximum flood currents** and **ebb currents** and the times of **slack water** to be encountered in numerous coastal waterways.

curvature effect—For small liquid water droplets, the increase in **equilibrium vapor pressure** due to the curved water surface; this may inhibit droplet growth.

cutoff high—A **warm high** (**anticyclone**) that has become displaced out of the basic westerly current and lies to the north of this current. Frequently, such highs are also **blocking highs**.

cutoff low—A **cold low** (**cyclone**) that has become displaced out of the basic westerly current and lies to the south of this current.

cutting-off process—A sequence of events by which a **warm high** or **cold low**, originally within the westerlies, becomes displaced either poleward (**cutoff high**) or equatorward (**cutoff low**) out of the westerly current. This process is evident at very high levels in the atmosphere; and it frequently produces, or is part of the production of, a **blocking** situation.

cycle—Any series of recurring events that repeat with some sense of periodicity; also refers to one such complete sequence of recurring events and the associated time interval required for its completion.

cyclogenesis—Any development or strengthening of **cyclonic** circulation in the atmosphere. Opposite of **cyclolysis**. Cyclogenesis is applied to the development of cyclonic circulation where previously it did not exist (commonly, the initial appearance of a **low** or **trough**), as well as to the intensification of existing cyclonic flow. While cyclogenesis usually occurs together with **deepening** (a decrease in atmospheric pressure), the two terms should not be used synonymously.

cyclolysis—Any weakening of **cyclonic** circulation in the atmosphere. Opposite of **cyclogenesis**. Cyclolysis, which refers to the circulation, is to be distinguished from **filling**, an increase in atmospheric pressure, although the two processes commonly occur simultaneously.

cyclone—(1) A weather system characterized by relatively low surface air pressure at a given level and a closed **cyclonic** wind circulation; essentially the same as a **low**. A cyclone's direction of rotation (counterclockwise in the North Hemisphere) is opposite to that of an **anticyclone**. (2) Regional name for a severe **tropical cyclone** with sustained surface wind speeds greater than 64 knots (74 mph) in the western South Pacific and Indian Oceans, equivalent of a **hurricane** or **typhoon**.

cyclone family—A series of **wave cyclones** occurring in the interval between two successive major **polar outbreaks** of polar air. The series travels along the **polar**

front, usually eastward and poleward. Typically, the polar front drifts eastward and equatorward, so that each new cyclone of the family has its origin and trajectory at a lower latitude than the previous cyclone.

cyclone wave—(1) A disturbance in the lower **troposphere**, of wavelength 1000–2500 kilometers (*synoptic scale*). They are recognized on synoptic charts as migratory high and low pressure systems. These waves have been identified with the unstable perturbations discussed in connection with **baroclinic instability**. (2) A frontal wave with a crest containing a center of cyclonic circulation; therefore, the frontal wave of a **wave cyclone**.

cyclonic—Having a sense of rotation about the local vertical the same as that of the earth's rotation—that is, as viewed from above, counterclockwise in the Northern Hemisphere, clockwise in the Southern Hemisphere, undefined at the equator. Opposite of **anticyclonic**.

cyclonic shear—Horizontal **wind shear** of such a nature that it contributes to the cyclonic **vorticity** of the flow; that is, it tends to produce cyclonic rotation of the individual air particles along the line of flow. In the Northern Hemisphere, cyclonic shear is present if (when one faces downwind) the **wind speed** increases from left to right across the direction of flow; the opposite is true in the Southern Hemisphere. Contrast with **anticyclonic shear**.

D

D-layer—The lowest "layer" of the **ionosphere**. It exists only in the daytime. It is not strictly a layer at all because it does not exhibit a peak of electron or ion density, but is rather a region of increasing electron and ion density, starting at about 70–80 kilometer altitude and merging with the bottom of the **E-layer**. The layer attenuates AM radio waves.

daily maximum temperature—Maximum temperature during a continuous time interval of 24 hours, usually local midnight to midnight.

daily mean—(1) The average value of a **weather element** (e.g., temperature, precipitation, humidity) over a period of 24 hours. The "true daily mean" is usually taken as the mean of 24 hourly values between midnight and midnight, either as continuous values taken from an autographic record or as point readings at hourly intervals. When hourly values are not available, approximations must be made from observations at fixed hours. For temperature, a number of formulas have been devised for such approximations. Often simply the arithmetic mean of the daily maximum and minimum is used. For other elements the mean of the available observations is usually accepted. (2) The long-period mean value of a **climatic element** on a given day of the year.

daily minimum temperature—Minimum temperature during a continuous time interval of 24 hours, usually local midnight to midnight.

Dalton's law—The empirical generalization that for many so-called **perfect gases**, a mixture of these gases will have a pressure equal to the sum of the **partial pressures** that each of gas species would have as sole component with the same whole volume and temperature, provided no chemical interaction occurs. Named for John Dalton (1766–1844), a British chemist who formulated the concept.

damped wave—Any wave phenomenon whose **amplitude** decreases with time or whose total energy decreases by transfer to other portions of the wave spectrum.

damping—The decrease, with time, in **amplitude** or energy of atmospheric disturbances, owing either to a reversible **energy conversion** process or to the action of viscous or frictional effects.

dangerous semicircle—That half of a **tropical cyclone** to the right of the direction of movement of the storm in the Northern Hemisphere (to the left in the Southern Hemisphere), the other half being termed the *navigable semicircle*.

dart leader—The **leader** that typically, after the first **stroke**, initiates each succeeding stroke of a **composite flash** of lightning. (The first stroke is initiated by a **stepped leader**.)

data assimilation—The procedure whereby data, from various sources and usually different types, are combined to produce a dataset that is consistent both horizontally and vertically.

Davidson Current—A **countercurrent** of the northeastern Pacific Ocean running north along the west coast of the United States (from northern California to Washington to at least 48°N) during the winter months.

dawn—The first appearance of light in the eastern sky before **sunrise**; or the time of that appearance. It is synonymous with ''daybreak'' and the beginning of morning **twilight**. The glow of dawn is produced by the scattering of light reaching the upper atmosphere prior to the sun's rise to the observer's horizon.

day—A basic time increment defined by the earth's motion; specifically, a complete rotation of the earth about its own axis. The **sidereal day** is defined as the time required for the earth to make one complete rotation in an absolute coordinate system—that is, with respect to the stars. The day in common use is the **mean solar day**, derived, by means of the **equation of time**, from the **apparent solar day**, which is determined directly from the apparent relative motion of the sun and earth. The *civil day* is a modification of the mean solar day, which renders it practical as a time measure for ordinary purposes.

dead reckoning—A method of navigation utilizing only the **speed** and **heading** of the craft, without reference to external aids.

debris cloud—A visible aggregate of dust or debris observed to be rotating beneath a **funnel cloud** or around the base of a **tornado**. The debris cloud may extend several meters above the earth's surface and may provide an indication of the presence of a tornado. Compare with **dust whirl**.

decibel (dB)—(1) A quantity used to compare power levels, defined as 10 times the common logarithm of the ratio of the power levels. (2) A standard unit of sound volume measurement based on a logarithmic ratio of sound intensity and a standard sound.

declination—(1) In astronomy, the angular distance (in degrees) between any celestial object (e.g., sun, planets, satellites) and the **celestial equator**, measured positive to north and negative to south along a **great circle** passing through the object and the **celestial poles**. See **solar declination**. (2) In terrestrial magnetism: at any given location, the angle between the geographical **meridian** and the **magnetic meridian**; that is, the angle between **true north** and **magnetic north**. Declination is either ''east'' or ''west'' according to whether the compass needle points to the east or west of the geographical meridian.

deep—An exceptionally deep depression in the ocean bottom, usually exceeding 6000-meter depth.

deepening—A decrease in the **central pressure** of a pressure system on a **constant-height chart**, or an analogous decrease in height (specifically, **geopotential height**) on a **constant-pressure chart**. Opposite of **filling**. The term is usually applied to a **low** rather than to a **high**, although technically it is acceptable in either sense.

deep water—The **water mass** normally found above **bottom water** and below the **thermocline**.

deep-water wave—Regular oscillations of the ocean surface in water that is deeper than one-half the average wavelength; its speed is independent of water depth; a **short wave** in oceanography. Contrast with **shallow-water wave**.

deflation—In geology, the lifting and removal of loose soil and other surface material by the wind, leaving the rocks bare to the continuous attack of the weather.

deformation—The change in shape of a **fluid** mass by space variations in the velocity field, specifically by stretching or shearing.

deformation-type thermometer—A thermal sensor (**thermometer**) with a sensing element that bends or deforms by an amount that is a function of temperature. A **bimetallic thermometer** is a deformation-type thermometer often used in a **thermograph**.

degree (°)—(1) A unit of **temperature**, describing an interval on a **temperature scale**, such as the **Fahrenheit** and **Celsius temperature scales**. (2) A unit of angular distance. See **radian**. The circumference of a circle contains 360°.

degree-day unit—Generally, a measure of the departure of the mean daily temperature from a given standard: one degree-day unit for each degree (°C or °F) of departure above (or below) the standard during one day. **Heating** and **cooling degree-day units** are the common examples.

de-icing—The process for removing unwanted ice formation from an object, such as an aircraft, by the application of appropriate chemicals, by heating, or by mechanical means.

dendritic crystal—A crystal, particularly an **ice crystal**, whose macroscopic form is characterized by intricate branching structures of a tree-like nature. Dendritic ice crystals possess hexagonal symmetry in their ideal forms and tend to develop when a crystal grows by **deposition** at temperatures between –16° and –12°C (3° and 10°F).

dendroclimatology—Study of climatic fluctuations using the annual growth rings of certain trees.

dense fog advisory—Issued by the National Weather Service to caution the public of the possibility that the visibility could be reduced by dense **fog** to 0.25 mile (0.4 kilometer) or less.

density—Usually refers to the *mass density*, or the ratio of the mass of any substance to the volume occupied by it (usually expressed in kilograms per cubic meter, but any other unit system may be used); the reciprocal of **specific volume**. See **number density**.

density altitude—The altitude in the **standard atmosphere** at which the air has the same density as the air at the point in question. The **pressure altitude** corrected for temperature departures from the standard atmosphere. Useful for assessing the performance characteristics of aircraft.

density correction—That part of the **temperature correction** of a **mercury barometer** that is necessitated by the variation of the density of mercury with temperature.

density current—The intrusion of a denser **fluid** beneath a lighter fluid, due mainly to the hydrostatic forces arising from **gravity** and the density differences. Examples include **fall winds** and **salinity currents**.

deposition—Process by which water changes phase directly from a vapor into a solid (ice) without first becoming a liquid; an example is **frost formation**. Opposite of **sublimation. Latent heat** of deposition is released by the water molecules into the system.

depression—In meteorology, an area of low pressure at any given level in the atmosphere; a **low** or **trough**. This term is usually applied to a certain stage in the development of a **tropical cyclone**, to migratory lows and troughs, that are only weakly developed or poorly organized. See **tropical depression**.

depression storage—Water retained in puddles, ditches, and other depressions in the surface of the ground; the small-scale counterpart of **closed drainage**. In hydrology, the water volume needed to fill small depressions to overflow levels.

depth—In general, the vertical distance, usually directed from top to bottom. (1) In oceanography, the vertical distance from below the sea surface, or **mean sea level** at any given point, especially on the seafloor. (2) In hydrology, the vertical measure of any water body.

depth hoar—Ice crystals formed within snow below the surface as a result of **deposition**; a type of **hoarfrost**.

depth of no motion—In oceanography, the **depth** where water is assumed to be motionless; this level is used as a reference surface for calculating **geostrophic currents**. See **layer of no motion**.

derecho—A widespread convective windstorm consisting of a complex of **thunderstorms** that develop into a long-lived **squall line** or a **Mesoscale Convective Complex**, with straight-line winds. Literally, "straight ahead" to differentiate straight-line winds from wind vortices in **tornadoes**. Now considered as any family of **downburst** winds moving primarily from northwest to southeast. The area affected may extend for hundreds of kilometers along the path of the system.

desert—A region where precipitation is insufficient to support any except xerophilous vegetation; a region of extreme aridity.

desert climate—A **climate** type that is characterized by insufficient moisture to support appreciable plant life—that is, a climate of extreme aridity. See also **arid climate**.

desert wind—A wind blowing off the **desert**. It is very dry and usually dusty, very hot in summer but cold in winter, with a large diurnal range of temperature; an example is the **harmattan**.

desertification—The formation or increase of desertlike conditions in a region as a result of a permanent lack of water within a region from decreased rainfall, a lack of irrigation or possibly deforestation or overcropping.

desiccation—In climatology, the permanent decrease or disappearance of water from a region. This may be due to (*a*) a decrease of rainfall, (*b*) a failure to maintain irrigation, or (*c*) deforestation or overcropping; this process may be the result of natural or anthropogenic causes. Contrast with **exsiccation**.

development—The process of intensification of an atmospheric disturbance, most commonly applied to **cyclones** and **anticyclones**.

dew—Liquid water condensed directly onto grass and other objects near the ground from water vapor present in the atmosphere; dew is not considered to be **precipitation**. Dew occurs when the temperatures of the surface objects have fallen below the **dewpoint** of the ambient surface air either because of **radiational cooling**, especially at night, or because of the **advection** of moist air with a dewpoint higher than the local surface temperature. In either case, temperatures when saturation and condensation occur must be above freezing. Contrast with **frost**. Dew that freezes after formation is called **white dew**.

dew cell—An instrument used to determine the **dewpoint**. It consists of a pair of spaced bare electrical wires wound spirally around an insulator and covered with a wicking wetted with a water solution containing an excess of lithium chloride. An electrical potential applied to the wires causes a flow of current through the lithium chloride solution, which raises the temperature of the solution until its **vapor pressure** is in **equilibrium** with that of the ambient air. This temperature is measured by a **thermistor** and represents the dewpoint.

dewpoint—The temperature to which a given **air parcel** must be cooled at constant pressure and constant water vapor content in order for **saturation** to occur. See **frost point**.

dewpoint depression—The difference in degrees between the air temperature and the **dewpoint**.

dewpoint hygrometer—An instrument for determining the **dewpoint**; a type of **hygrometer**. It operates in the following manner. An **air parcel** is cooled at constant pressure, usually by contact with a refrigerated polished metal surface. Condensation as a thin foggy film appears upon the metal surface at a temperature slightly below that of the thermodynamic dewpoint of the air. The observed dewpoint will differ from the thermodynamic dewpoint depending upon the nature of the condensing surface, the condensation nuclei, and the sensitivity of the condensate-detecting apparatus.

diabatic process—Thermodynamic **change of state** of a system in which transfer of heat occurs across the boundaries of the system. This usage is to be preferred to "nonadiabatic."

diffluence—The rate at which adjacent flow is diverging along an axis oriented normal to the flow at the point in question. Opposite of **confluence**. Progressive drawing apart of the **streamlines** in the direction of flow. This situation may cause **divergence**.

diffraction—The interference patterns that are produced when a slight bending of a light wave moving along the boundary of an object such as a water droplet causes the waves to become superimposed; the constructive and destructive interference of the light produces light, dark, or colored bands, such as **corona** (1) and **glory** effects.

diffuse front—A **front** across which the characteristics of **wind shift** and temperature change are only weakly defined.

diffuse radiation—Any **radiant energy** propagating in many different directions through a given small volume of space.

diffuse reflection—(1) The change in the direction of radiation in all directions from a rough surface. Contrast with **specular reflection**. (2) Term frequently applied to the process by which **solar radiation** is scattered by dust and other suspended particles in the atmosphere.

diffuse sky radiation—The component of **solar radiation** reaching the earth's surface after having been **scattered** from the direct solar beam by molecules or aerosols in the atmosphere. Of the total light removed from the direct solar beam by scattering in the atmosphere, about two-thirds ultimately reaches the earth as diffuse sky radiation.

diffusion—The exchange of **fluid** parcels, and, hence, the **transport** and spread of **conservative properties** between regions in space by molecular motions or by eddy motions. Diffusion is always directed from higher toward lower concentrations. In most meteorological applications, these processes (where turbulence predominates) have apparently random motions on a scale too small to be treated by the ordinary **equations of motion**. See **turbulent diffusion**.

dimensional analysis—An empirical diagnostic method of examination of physical systems that depends only on the dimensions of the physical variables involved.

Dines' compensation—A model developed by William H. Dines (1855–1927), a British meteorologist, to link the horizontal and vertical motions in the **troposphere** and to explain the maintenance of **synoptic-scale** atmospheric pressure systems. Regions of low-level horizontal **convergence** in low pressure systems are compensated in the upper troposphere by horizontal **divergence** aloft, while surface divergence in high pressure systems are maintained by compensating upper-level convergence, above the **level of nondivergence**.

dipole—A system composed of two equal electric or magnetic charges of opposite sign separated by a finite distance. It frequently refers to a molecule in which the centers of positive and negative electric charges are separated, thereby giving the molecules an electric moment (**dipole moment**). The water molecule (H_2O) is a dipolar molecule.

dipole moment—A **vector** quantity representing the product of the strength of the electric charge and the separating distance produced by a **dipole**. The water vapor molecule is the only one among the atmospheric gases with a permanent dipole moment. Thus, many characteristics of the atmosphere as a conductor of electromagnetic radiation vary with humidity.

direct cell (or *direct circulation*)—A closed thermal circulation in a vertical plane where warm air rises and cold air sinks (or equivalently, a regime in which the rising motion occurs at a higher potential temperature than the sinking motion). Such a cell converts **heat energy** to **potential energy** and then to **kinetic energy**. An example of such a cell in the atmosphere is the **subtropical** and **equatorial** circulation. Contrast with **indirect circulation**.

direct solar radiation (also known as *beam radiation*)—The component of **solar radiation** that is transmitted to the earth's surface only from the direction of the sun's disk, and, as a beam, would cast a shadow. This excludes **diffuse sky radiation** from clouds or clear sky in any other direction.

discharge—Volume of water flowing through a cross section of a waterway per unit time (usually cubic feet per second, **cfs**). The discharge often is specified either as *"mean discharge,"* defined as the arithmetic average of all daily mean discharges for a specific time interval, or the *"instantaneous discharge,"* the observed discharge at a particular time.

discontinuity—The abrupt variation or sudden change of a variable at a line or surface. A **front** often is considered to be a discontinuity.

dish—A parabolic reflector type of radar or radio antenna; often used in referring to all types of radar antennae or many satellite receivers.

dispersion—Any separation of a complex wave into its components. Generally refers to the process in which **radiation** (e.g., **polychromatic** sunlight) is separated into its separate wavelengths (or component colors). Dispersion results when an optical process, such as **diffraction**, **refraction**, or **scattering**, varies according to wavelength.

disphotic zone—A layer within a large body of water, usually ranging from 100 to 600 meters below the surface that does not have sufficient sunlight for photosynthesis. Contrast with the **euphotic zone**.

display—(1) Any graphical presentation of data. (2) In radar, the array of light and dark or colored areas on the face of the **radar screen** representing the spatial distribution of echoes identifying **targets**.

dissipation trail—A clear rift left behind an aircraft as it flies in a thin cloud layer. Opposite of a **condensation trail**.

distribution graph—In hydrology, a modified unit **hydrograph** that is usually presented as a histogram or table of percent **runoff** within each of a series of successive short time intervals.

disturbance—In general, any agitation or disruption of a steady state. In meteorology, this has several rather loose applications: (a) used for any **low** or **cyclone**, but usually one that is relatively small in size and effect; (b) applied to an area where weather, wind, pressure, etc., show signs of the development of cyclonic circulation (see **tropical cyclone**); (c) used for any deviation in flow or pressure that is associated with a disturbed state of the weather—that is, cloudiness and precipitation; and (d) applied to any individual circulatory system within the primary circulation of the atmosphere (see **tropical disturbance**).

diurnal—Daily, especially pertaining to periodic-type actions that are completed within 24 hours and that recur every 24 hours; thus, most reference is made to diurnal cycles, variations, ranges, maxima, etc. Compare with **annual**.

diurnal inequality (or *daily inequality*)—In oceanography, the difference between the heights of the two successive **high waters** or the two **low waters** of a **lunar day**; also the difference in velocity (speed and direction) of the two **flood currents** or two **ebb currents** of each day.

diurnal tide—A tide with only one **high water** and one **low water** occurring each **lunar day**.

divergence—A **fluid** flow pattern producing a net outflow of fluid elements from a central point. In mathematical discussion, divergence is taken to include **convergence**— that is, negative divergence. In meteorology, because of the predominance of horizontal motions, the divergence usually refers to the two-dimensional horizontal divergence of the velocity field. Horizontal divergence is accompanied by compensating vertical motion. Compare with **diffluence**.

Dobson spectrophotometer—A calibrated instrument designed by G.M.B. Dobson to measure the **ozone** content of the atmosphere by measuring the incoming solar **ultraviolet** radiation absorbed by ozone. The instrument measures and compares the relative radiation intensity at two different ultraviolet wavelengths.

Dobson unit—The depth of **ozone** (O_3) produced if all of the ozone in a column of the atmosphere were brought to sea level temperature and pressure; one Dobson unit is a hundredth of a millimeter. Named for G.M.B. Dobson, who made extensive ozone measurements with his **Dobson spectrophotometer**.

dog days—Colloquially, the period of greatest heat in summer; typically occurring between late July and early September for residents of midlatitudes in the Northern Hemisphere.

doldrums—A nautical term for the **equatorial trough**, with special reference to the light and variable nature of the winds. This region typically is within 5°–10° of the equator.

Doppler effect—The shift in frequency of an **electromagnetic** or **sound wave** due to the relative movement of the source or the observer; the amount of shift depends upon this relative speed. Named for Christian J. Doppler (1803–1853), an Austrian physicist who explained the phenomenon.

Doppler radar—A **weather radar** system distinguished from conventional units because it operates on the **Doppler effect**. It has the capability to determine the **radial velocity** of **targets** (precipitation, dust particles) moving directly toward or away from the radar unit based on the difference in frequency (or phase) between the outgoing and returning radar signal.

dosimeter—(1) An instrument for measuring the ultraviolet radiation in solar and sky radiation. (2) An instrument for detecting a person's cumulative exposure (dosage) to radiation.

downburst—An exceptionally energetic **downdraft** that exits the base of a **thunder-**

storm (not always severe) and spreads out at the earth's surface as strong and gusty horizontal winds that may cause property damage. See **microburst** and **macroburst**. It may be accompanied by precipitation.

downdraft—A term applied to the strong downward-flowing air current within a **thunderstorm,** which is usually associated with **precipitation**. An exceptionally strong downdraft can result in a **downburst**.

downstream—The direction toward which a **fluid** is moving, usually implying the horizontal component of the mean direction or direction of the basic current. Opposite of **upstream**.

downwelling—(1) The component of radiation (either solar or atmospheric) directed downward toward the earth's surface. (2) The **sinking** of water away from the surface toward subsurface layers because of surface **convergence**. Opposite of **upwelling**.

downwind—The direction toward which the wind is blowing; with the wind. Opposite of **upwind**.

drag—The frictional impedance offered by a **fluid** to the motion of bodies passing through it; in more precise meteorological usage, the component of **aerodynamic** force parallel to the direction of mean flow.

drainage—Removal of **surface water** or **groundwater** from a region by **gravity** or by pumping.

drainage area—The size of the area of a **watershed** or **river basin**.

drainage basin—Any area having a common outlet for surface **runoff**.

drainage density—One of the means of numerically characterizing a stream for study and comparative purposes, computed by dividing the total length of the stream **channel** by its **drainage area**; usually expressed in miles per square mile.

drift—(1) In general, any deviation from a course caused by wind or current. The effect of the velocity of a **fluid** flow upon the velocity (relative to a fixed external point) of an object moving within the fluid; the vector difference between the velocity of the object relative to the fluid and its velocity relative to the fixed reference. (2) The speed of an **ocean current**. In publications for the mariner, drifts are usually given in miles per day or in knots. (3) Any rock material (e.g., boulders, till) transported by a **glacier** and deposited by the ice or **meltwater**.

drift bottle—A bottle that is released into the sea for use in studying **surface currents**. It contains a card, identifying the date and place of release, to be returned by the finder with the date and place of recovery.

drift current—Same as **wind-drift current**.

drift ice—Any **sea ice** that has traveled (drifted) from its place of origin. The term is used in a wide sense to include any areas of sea ice, other than **fast ice**, no matter what form it takes or how disposed.

drifting snow—An ensemble of snow particles raised by the wind to a small height above the ground. The **visibility** at eye level (6 feet) is not significantly reduced below 7 miles (10 kilometers). Contrast with **blowing snow**, where visibility is reduced.

drizzle—A type of **liquid precipitation** composed of very small, numerous, and uniformly dispersed water drops that may appear to float while following air currents; consists of water drops of diameter 0.2–0.5 millimeters (0.01–0.02 inch) falling through the atmosphere. Unlike **fog** droplets, drizzle falls to the ground. It usually falls from low **stratus** clouds and is frequently accompanied by low **visibility** and fog.

drogue—A sea anchor or other parachute-shaped device for use in water. Drogues suspended at desired depths by buoys are used to determine the **set** and **drift** of currents at those depths by following the motions that they give to the buoys at the surface.

droplet—A small spherical particle of any **liquid**; in meteorology, particularly a water droplet. No defined size limit separates droplets from drops of water.

dropsonde—A **radiosonde** that is dropped by parachute from an aircraft for the purpose of obtaining temperature, pressure, and humidity profiles of the atmosphere below flight level; often used by aircraft reconnaissance of **hurricanes**.

drought—An extended interval of abnormally dry weather sufficiently prolonged for the lack of water to cause a serious hydrologic imbalance (i.e., crop damage, water supply shortage, etc.) in the affected area; typically more extensive and longer than a **dry spell**.

drought index—A numeric value related to some of the cumulative effects of prolonged and abnormal moisture deficiency. In the United States, the *Palmer Drought Severity Index* is widely used by agricultural and other interests; values of this index are computed biweekly by the National Weather Service during the **growing season** for each climatological division in the United States.

dry adiabat—On a **thermodynamic diagram**, a line that represents the **dry-adiabatic lapse rate,** describing the temperature change experienced by an unsaturated **air parcel** in an **adiabatic process** as the parcel is moved vertically in the atmosphere and undergoes expansional cooling during ascent and compressional heating on descent. It also represents a line of constant **potential temperature**.

dry-adiabatic lapse rate—A special **process lapse rate** of temperature, defined as the rate of decrease of temperature with altitude of a **dry** (unsaturated) **air parcel** lifted adiabatically through an atmosphere in **hydrostatic equilibrium**. This lapse rate is numerically equal to 9.767°C per kilometer or about 5.4°F per thousand feet.

dry air—In atmospheric thermodynamics and chemistry, air that contains no **water vapor**. Popularly refers to air of low **relative humidity**.

dry-bulb temperature—Technically, the temperature registered by the dry-bulb thermometer of a **psychrometer**. However, it is identical with the **ambient air temperature** and may also be used in that sense. Contrast with **wet-bulb temperature**.

dry deposition—Removal of suspended particles from the atmosphere through **impaction** and **gravitational settling**; a natural means whereby air is cleansed of **pollutants**.

dry fog—(1) A fog **that** does not moisten exposed surfaces. (2) A condition of reduced visibility due to the presence of **dust**, **smoke**, or **haze** in the air; it is not a true fog. Contrast with **smog**.

dryline—A nonfrontal boundary between warm, dry air and warm, **humid air**; a likely site for thunderstorm development; often observed in west Texas in spring.

dry season—In certain types of **climate**, an annually recurring period of one or more months during which precipitation is at a minimum for that region. Opposite of **rainy season**.

dry spell—Loosely, an interval of abnormally dry weather. The term should be reserved for a less extensive, and therefore less severe, condition than a **drought**. For many locales in the United States, an interval of at least 2 weeks without measurable precipitation—that is, greater than a **trace**.

dry tongue—In **synoptic meteorology**, a pronounced protrusion of relatively dry air into a region of higher humidity.

dust—Fine, solid earthen particles (soil, dust, or sand) suspended in the air by **duststorms** or **sandstorms**. Because of the scattering of light, dust gives a tan or gray image to distant objects—for example, the solar disk appears pale and colorless. Compare with **haze** and **smoke**.

Dust Bowl—A name given, early in 1935, to the region in the south-central United States that at that time was afflicted by **drought** and **duststorms**. It included parts of five states: Colorado, Kansas, New Mexico, Texas, and Oklahoma. It resulted from a long period of deficient rainfall combined with loosening of the soil by destruction of the natural vegetation. The name has since been extended to similar regions in other parts of the world.

dust devil—A well-developed **dust whirl**; a small but vigorous **whirlwind**, usually of short duration, rendered visible by **dust**, **sand**, or debris picked up from the ground. Diameters range from about 3 meters (10 feet) to greater than 30 meters (100 feet); their average height is about 200 meters (600 feet), but a few have been observed as high as a thousand meters. Dust devils typically occur over bare ground under intense **insolation**.

duststorm—An unusual, frequently severe weather condition characterized by strong winds and dust-filled air over an extensive area. Prerequisite to a duststorm is a period of **drought** over an area of normally arable land, thus providing the very fine particles of dust that distinguish it from the more common **sandstorm** of desert regions. In accordance with United States observational practice, a *duststorm* is reported if the horizontal visibility due to blowing dust were reduced to less than 1 kilometer (5/8 mile), but not less than 500 meters (5/16 mile); a *severe duststorm* is reported if the visibility due to dust were to become less than 500 meter (5/16 mile).

duststorm warning—Issued by the National Weather Service to warn the public that **blowing dust** and/or **blowing sand** will reduce visibility to near zero. Contrast with **blowing dust/blowing sand advisory.**

dust whirl—A rapidly rotating column of air (**whirlwind**) over a dry and dusty or sandy area, carrying dust, leaves, and other light material picked up from the ground. If it also contains debris, it would be called a **debris cloud**. When well developed it is known as a **dust devil**, or in the extreme, it may be called a **gustnado**.

dynamic climatology—The statistical collation and study of observed elements (or derived parameters) of the atmosphere, particularly in relation to the physical and dynamical explanation or interpretation either of the contemporary climate patterns with their anomalous fluctuations or of the long-term **climate changes** or **trends**.

dynamic height (or ***geopotential height***)—Used in oceanography to describe the amount of **work** done when a water parcel of unit mass is moved vertically between two levels. Dimensions are potential energy per unit mass. See **dynamic meter**.

dynamic meteorology—The study of atmospheric motions as solutions of the fundamental equations of **hydrodynamics** or other systems of equations appropriate to special situations, as in the statistical theory of turbulence.

dynamic meter—The standard unit of **dynamic height** in oceanography; expressed as 10 square meters per second squared.

dynamic pressure—The **pressure** exerted by a **fluid** either by virtue of the motion of the fluid against an object or by the motion of the object through the fluid. Represents the difference between the **static pressure** and the pressure measured by a **barometer** moving relative to the fluid.

dynamic trough—A pressure **trough** formed on the lee side of a mountain range across which the wind is blowing almost at right angles; often seen on United States weather maps east of the Rocky Mountains, and sometimes east of the Appalachians, where it is less pronounced.

dynamics—The study of the action of forces on bodies and the changes in motion these forces produce. Compare with **kinematics**.

dynamical system—A system such as the earth–atmosphere whose evolution from some initial state in time can be described by one or more mathematical equations.

E

E-layer—The "layer" of the **ionosphere** that is usually found at an altitude between 100 and 120 kilometers. It exhibits one or more distinct maxima and sharp gradients of **free-electron** density. It is most pronounced during daytime but does not entirely disappear at night.

earth current—A large-scale surge of electric charge within the earth's crust associated with a disturbance of the **ionosphere**.

earth mound (or *earth hummock*)—A small dome-shaped uplift of soil caused by the pressure of **groundwater**. Compare with **frost mound**.

earth radiation—Same as **terrestrial radiation**.

earth shadow—The shadow of the earth in interplanetary space; it has the shape of a right cone and consists of an *umbra* (central dark part of shadow) and *penumbra* (outer part of shadow) into which the moon must pass to produce a lunar eclipse.

earthlight (or *earthshine*)—The faint illumination of the dark part of the moon's disk produced by sunlight reflected onto the moon from the earth's surface and atmosphere.

East Australian Current—A broad southward-flowing ocean current in the South Pacific that parallels the east coast of Australia between 20° and 45°S, before turning northward along the west coast of the South Island of New Zealand.

East Greenland Current—A cold ocean current in the North Atlantic flowing southward and southwestward along the east coast of Greenland. It represents the main outlet from the Arctic Ocean, with a majority of water passing through the Denmark Strait between Iceland and Greenland. This current rounds the southern tip of Greenland, contributing to the **West Greenland Current**. The rest of the water flows **cyclonically** into the Norwegian Sea.

easterlies—Any winds with components from the east, usually applied to broad air currents or patterns of persistent easterly winds, the "easterly belts," such as the **equatorial easterlies**, the **tropical easterlies** (see also **trade winds**), and the **polar easterlies**. Contrast with **westerlies**.

easterly wave—A migratory wavelike disturbance of the **tropical easterlies** that typically moves from east to west; a **hurricane** may form as an easterly wave. Contrast with **westerly wave**.

ebb current—The movement of a **tidal current** away from the coast or down an **estuary** or tidal waterway. Opposite of **flood current**.

ebb tide—Same as **falling tide**.

eccentricity (or *orbital eccentricity*)—A dimensionless quantity describing the elliptical shape of a planetary orbit, with a larger value, indicating a more elliptical orbit. A factor in the **Milankovitch cycles**, in that it directly influences the sun–earth distance at **perihelion** and **aphelion** and contributes to the variation in seasonal length. Presently, the eccentricity of the earth's orbit is 0.0167 and it undergoes a highly complex periodic variation ranging from 0.001 (nearly circular) to 0.05 (more elliptical), with dominant periodic components of 100 000 and 400 000 years.

echo—In **radar meteorology**, a general term for the appearance, on a **radar screen**, of the microwave energy reflected or scattered back from a **target**. A **false echo** may be detected because of **anomalous propagation** or nonprecipitation targets.

echo intensity—The relative strength of a **radar echo** as displayed on a **radar screen**; used to estimate the rate and cumulative amount of precipitation, often calibrated against a color scale. See **VIP level**.

echo signal—Same as **target signal**; the reflected or scattered energy returned to a radar unit by a **target** and seen as an **echo** on the **radar screen**.

ecliptic—The mean plane of the earth's annual orbit about the sun; the intersection of this orbital plane with the **celestial sphere** produces a **great circle** on the celestial sphere, which describes the apparent annual path of the sun across the sky.

ECMWF—See **European Centre for Medium-Range Weather Forecasts**.

ecological climatology (or *ecoclimatology*)—A branch of **bioclimatology** that studies the relationships between organisms and their climatic environment. It includes the physiological adaptation of plants and animals to their climate and the geographical distribution of plants and animals.

ecology—The study of the mutual relations between organisms and their environment.

eddy—(1) Irregular whirls within a **fluid**, such as air or water, that have a definable volume and lifetime and carry the properties of the fluid. (2) The rotational movement occurring in a flowing fluid.

eddy diffusion—Same as **turbulent diffusion**.

edge wave—An ocean wave traveling parallel to a coast, with crests perpendicular to the coastline. Such a wave has a height that diminishes rapidly seaward and is negligible at a distance of one wavelength offshore.

effective precipitation—(1) That part of precipitation that reaches stream channels as **runoff**. (2) In irrigation, that portion of the precipitation that remains in the soil and is available for **consumptive use**.

effective temperature—(1) An engineering index, defined as the equilibrium air temperature within an enclosure with air at a **relative humidity** of 50% that would induce the same total **sensible** and **latent heat** exchange as that induced by the actual conditions of temperature, humidity, and air movement. (2) With respect to radiation, ascribed to an imperfectly radiating body; the temperature at which a perfect radiator (**blackbody**) would emit radiation at the same rate. Thus, the effective temperature is always less than the actual temperature.

effective terrestrial radiation—The difference between the outgoing **infrared** or **terrestrial radiation** emitted from the earth's surface and the downwelling infrared counter-radiation from the atmosphere. See **nocturnal radiation**.

Ekman layer (also called *spiral layer*)—Part of the **planetary boundary layer** lying between the **surface boundary layer** (approximately at **anemometer level**) and the **free atmosphere**, where the atmosphere is assumed an ideal fluid in **geostrophic balance**. Flow within this layer is described by the **Ekman spiral**; frequently termed **mixed layer**.

Ekman spiral—An idealized representation of how **wind-driven currents** in the surface layers of the ocean vary with depth. Named for Vagn Walfrid Ekman (1874–1954), a Swedish physicist. In the atmosphere, the variation of the horizontal wind with altitude within the **friction layer** or **planetary boundary layer**. This vertical wind variation appears as a helical-shaped envelope (spiral) on a **hodograph:** typically, an increase in wind speed with height and a turning of winds from a marked cross-isobar flow toward low pressure at the surface to a flow at the top of the layer tending to parallel the isobars in an essentially **geostrophic** flow aloft. In the Northern Hemisphere oceans, the near-surface **drift** is to the right of the wind, while in the Southern Hemisphere, drift is to the left.

El Niño—Anomalous warming of surface ocean waters in the eastern tropical Pacific; accompanied by suppression of upwelling off the coasts of Ecuador and northern Peru and heavy rainfall in the coastal regions of those nations. Because this condition often occurs around Christmas, it is named El Niño (Spanish for boy child, referring to the Christ child). In most years, the warming lasts only a few weeks or a month, after which the **sea surface temperatures** and weather patterns return to normal. However, when El Niño lasts for many months, more extensive ocean warming occurs and weather extremes occur in widely separated regions of the globe, possibly as a result of **teleconnections**. See **ENSO**. Contrast **La Niña**.

electric field—A region in which any charged particle would experience an electrical force. In meteorology, the term is used to describe the atmospheric "*electric field strength,*" or the magnitude of the electrical force exerted on a unit positive charge placed at a particular point the electric field (i.e., space).

electrical discharge—An **exothermic** process with a flow of electricity through a gas, resulting in the emission of **electromagnetic radiation**. Examples include **point**, **lightning,** and **corona discharges.**

electrical storm—Popular name for a **thunderstorm**, often with frequent **lightning**.

electrical-type thermometer—A thermal sensor (**thermometer**) instrument with a sensor element having electrical properties that are a function of temperature. See **thermocouple**, which utilizes a generated voltage potential, or **thermistor**, which utilizes electrical resistance.

electromagnetic radiation (or simply ***radiation***)—A form of **energy** propagated through space or through material media by advancing transverse wave disturbances in electric and magnetic fields existing in space or in the media. The speed of this propagation through space is approximately 3×10^8 meters per second (the classic "*speed of light*").

electromagnetic spectrum—The ordered continuum or array of all known types of **electromagnetic radiation**, which arranged according to **wavelength** extends from the shortest **cosmic rays** through **gamma rays**, **X-rays**, **ultraviolet radiation**, **visible radiation**, **infrared radiation**, **microwaves**, to the longest wavelengths of **radio energy**.

electrometeor—Any visible or audible indicator of **atmospheric electricity**, to include all types of **lightning discharges**, **thunder**, and **aurora**.

element—In general, one of the simplest parts of a complex entity, which has various uses: (1) *Chemical element*: A material that cannot be simplified further by ordinary chemical means; it consists of only one type of atom. (2) *Weather or climate element:* Any one of the observable properties or conditions of the atmosphere, which together specify the physical state of **weather** or **climate** at a given place for any particular moment or period of time (e.g., temperature, humidity, precipitation). (3) *Hydrological element:* Some hydrological phenomenon such as water level (**stage**), **discharge**, or **precipitation**.

elevation—Vertical distance of a point on or affixed to the surface of the ground, measured from **mean sea level**. Contrast with **altitude** (1).

elevation angle (also called ***altitude angle***)—The angle between the **horizon** and a point above the horizon in the **celestial sphere**, measured along the **great circle** that passes through the **zenith** and the point in question.

emission—(1) With respect to **electromagnetic radiation**, the generation and sending out of **radiant energy**. It is to be distinguished from **reflection** and **transmission** of incident radiation. (2) Discharge of gases or particles from a source such as a smokestack or exhaust pipe into the atmosphere, perhaps resulting in environmental **pollution**.

emissivity—The (dimensionless) ratio of the radiant energy **emittance** of a substance to the ideal emittance of an ideal (**blackbody**) radiator at that same temperature. This ratio may be applied to a particular wavelength or to the entire wavelength envelope.

emittance (or *exitance*)—A measure of the total **radiant energy** emitted per unit time per unit area of an emitting surface; the total **flux** of **electromagnetic radiation** emitted by a unit surface area. Units of emittance are watts per square meter. Compare with **radiance**, which is a directed radiant energy. Contrast with **irradiance** received on a surface.

endothermic—A process involving the absorption of heat **energy**. Contrast with **exothermic**.

energetics—The branch of study dealing with the systematic description of the **energy conversion** and **transfer** processes that take place within a physical system.

energy—The capacity for doing **work**; energy can be found in many forms, such as **kinetic**, **potential**, or **thermal energy**. In an **isolated system**, the total energy remains constant as a consequence of the **conservation of energy**, but may be converted from one form to another.

energy budget—A quantitative description or inventory of the **energy** exchange for a physical or ecological system to include specification of all energy fluxes that are sources and sinks of energy in the system; may be specified in an energy balance equation with each radiative and nonradiative flux identified.

energy cascade—Progressive transfer of **kinetic energy** from systems of larger scale to those of smaller scale, ending in its dissipation by **friction**.

energy conversion—The process whereby **energy** changes from one form to another (e.g., from **potential energy** to **kinetic energy**). It is common also to speak of *energy release* when referring to certain conversion processes, such as the release of **latent heat** during condensation of water vapor.

energy level—Any one of different **energy** states in which an atom may exist, where possible values are restricted to discrete or quantified conditions; the energy of the atom in different states of excitation.

energy transfer—The movement of **energy** of a given form among different scales of motion. For example, **kinetic energy** may be transferred between the zonal and meridional components of the wind, or between the mean and eddy components of the wind.

enhanced greenhouse effect—Intensification of the planet's natural **greenhouse effect** caused by an increase in the atmospheric concentrations of certain **greenhouse gases**; usually the increased concentrations are a consequence of various anthropogenic activities.

enhanced image—An **infrared satellite image** in which the apparent differences between certain adjacent tones or gray shades resulting from differences in emitted **infrared** (or **terrestrial**) **radiation** are artificially increased to reveal subtle temperature differences.

ensemble forecasting—A forecasting technique in which a numerical model generates several forecasts, each based on a slightly different set of **initial conditions**. If the forecasts are consistent, then the forecast would be considered reliable. If, however, the forecasts diverge, then any forecast would be considered unreliable.

ENSO (a contraction for **El Niño/Southern Oscillation**)—The term for the coupled ocean–atmosphere interactions in the tropical Pacific characterized by episodes of anomalously high **sea surface temperatures** in the equatorial and tropical eastern Pacific; associated with large-scale swings in surface air pressure between the western and eastern tropical Pacific; the most prominent source of interannual variability in weather and climate around the world. These episodes recur at irregularly spaced intervals (2–7 years) and may persist for as long as 2 years.

enthalpy—The **heat** content of a substance per unit mass. The change in enthalpy measures the heat imparted to a system during a reversible **isobaric process**. In meteorology, enthalpy is used in the context of **sensible heat** as opposed to **latent heat**.

entrainment—(1) In meteorology, the **mixing** of environmental air into a pre-existing organized air current so that the environmental air becomes part of the current. (2) In hydrology, the process of picking up and carrying away material from the bed or banks of a **channel**.

entropy—A mathematical concept based on the **second law of thermodynamics**. A measure of the **energy** of a system that has ceased to be available for **work** during a certain process.

environment—(1) External surroundings of a system. See **ambient**. (2) The complex of physical, chemical, and biological factors in which a living organism or community exists.

environmental lapse rate—The rate of decrease of air temperature with increasing altitude. This temperature **lapse rate** is a measurable quantity determined from a **sounding**, as compared with the theoretical **process lapse rate** involving thermodynamic changes in temperature attached to the vertical motions of an **air parcel** undergoing a particular adiabatic process over a specified distance.

eolation—The erosion of land surfaces by wind-driven dust or sand.

ephemeral lake *or* **ephemeral stream**—A lake or stream channel that carries water only during and immediately after periods of rainfall or snowmelt; it becomes dry during the **dry season** or a dry year. Contrast with **intermittent stream** and **perennial stream**.

epilimnion—The top layer of less dense water lying above the **thermocline** in a thermally stratified lake; analogous to a **mixed layer** (1) in an ocean. Contrast with **hypolimnion**.

equation of continuity—A mathematical statement of the **conservation of mass** in a **fluid**.

equation of state—A mathematical expression relating the **variables of state**, or specifically, the temperature, pressure, mass, and volume of a system in thermodynamic equilibrium: the state of an **ideal gas** can be specified as $p\alpha=RT$, where p is the pressure; α is the **specific volume**; R is the specific **gas constant**; and T is the absolute temperature.

equation of time (or more correctly, *equation of ephemeris time*)—The difference at any instant between the **apparent solar time** and the **mean solar time** as measured on any specified day of the year. Because of differences between the motions of the apparent (true) and mean (fictitious) sun associated with orbital **eccentricity** and **obliquity of the ecliptic**, the variation in the equation of time throughout the year is a complex periodic function of the day of the year, reaching as much as a 15-minute lead and lag time.

equations of motion—A set of mathematical statements representing the application of Newton's second law of motion (see **Newton's laws of motion**) to a **fluid** system. The total **acceleration** on an individual fluid particle is equated to the sum of the accelerations (or forces per unit mass) acting on the particle within the fluid.

equator—Geographically, the imaginary **great circle** of 0° latitude on the earth's surface, which is equidistant from the geographic **poles**, and which separates the Northern Hemisphere from the Southern Hemisphere. See **celestial equator**.

equatorial air—The air of the **doldrums** or the **equatorial trough**, to be distinguished somewhat vaguely from the **tropical air** of the **trade wind** zones. Tropical air "becomes" equatorial air when the former enters the equatorial zone and stagnates. No significant distinction exists between the physical properties of these two types of air in the lower troposphere.

equatorial climate—A **climatic zone** with a climate typical of regions along the equator. See **tropical climate**.

equatorial convergence zone—Same as **intertropical convergence zone (ITCZ)**.

Equatorial Countercurrent—An ocean current located within 10° of the equator and flowing eastward (counter to and between the westward-flowing **North Equatorial Current** and **South Equatorial Current**) through the Atlantic, Indian, and Pacific Oceans.

equatorial dry zone—An arid region existing in the **equatorial trough** resulting from low-level wind **divergence** and **subsidence**. The most famous dry zone is situated a little south of the equator in the central part of the equatorial Pacific.

equatorial easterlies—The **trade winds** (with an easterly component) in the summer hemisphere when they are very deep, extending to at least 8–10 kilometers altitude, and generally not topped by upper **westerlies**.

equatorial trough—The quasi-continuous belt of low atmospheric pressure lying between the **subtropical high-pressure belts** of the Northern and Southern Hemispheres. This entire region is one of very homogeneous air, probably the most ideally **barotropic** region of the atmosphere. Yet, atmospheric humidity is so high that slight variations in stability cause major variations in weather. This region typically is within 5°–10° of the equator, shifting slightly with the sun. Also known as the **doldrums**.

Equatorial Undercurrent (also known as ***Cromwell Current***)—An eastward-moving **subsurface current** flowing in the Pacific Ocean within 2° of the equator, extending from 150°E to 95°W.

equatorial westerlies—The westerly winds (blowing from the west) occasionally found in the **equatorial trough** and separated from the midlatitude **westerlies** by the broad belt of easterly **trade winds**.

equilibrium—In general, a state of balance within a system, where opposing forces or fluxes balance. In thermodynamics, any state of a system that would not undergo change if the system were to be isolated. Processes in an **isolated system** not in equilibrium are irreversible and always in the direction of equilibrium.

equilibrium vapor pressure—The **vapor pressure** of a system in which two or more phases of a substance (typically water) coexist in **equilibrium**.

equinoctial rains—**Rainy seasons** that occur regularly at or shortly after the **equinoxes** in many places within a few degrees of the equator.

equinoctial tide—Tide occurring when the sun is near **equinox**. During this period, **tidal ranges** in the **spring tide** are greater than average because of the orbital configuration.

equinox—(1) Either of two points on the sun's apparent annual path across the **celestial**

sphere where the sun is directly over the earth's geographic **equator**, such that the **ecliptic** and **celestial equator** coincide. (2) Popularly, the time at which the sun passes directly above the equator; the "time of the equinox." At this time essentially all locales experience 12 hours of sunlight. In the Northern Hemisphere, the **vernal equinox** falls on or about 21 March, and the **autumnal equinox** on or about 22 September. These dates are reversed in the Southern Hemisphere.

equivalent temperature—The temperature that an **air parcel** would have if all water vapor were condensed out at constant pressure, the **latent heat** of condensation released being used to heat the air.

erosion—The movement of soil or rock from one point to another by the action of the sea, running water, moving ice, precipitation, or wind. Erosion is distinct from **weathering**, for the latter does not necessarily imply **transport** of material.

escape velocity—The vertical speed that a particle in the **outer atmosphere**, especially in the **exosphere**, must attain in order to escape from the gravitational field of a planet or star.

estuarine circulation—A characteristic flow regime in an **estuary**, where the flow of water is seaward at the surface, but landward at depth.

estuary—The portion of a river affected by ocean **tides**; the semienclosed region in the vicinity of a river mouth, in which the **freshwater** of the river mixes with the **saltwater** of the sea.

etesian wind—A cool, dry, northeasterly wind found to prevail across the eastern Mediterranean and the Aegean Sea in summertime.

euphotic zone (also known as *photic zone* or *zone of photosynthesis*)—The near surface layer of a body of water (lake, river, ocean) that receives ample sunlight for the photosynthetic processes of plants. The layer is typically within the top 80 meters, but varies with sun angle and cloudiness. Contrast with **disphotic zone**.

Euroas—The Greek name for the stormy, rainy, southeast wind. On the **Tower of the Winds** at Athens this wind is represented by an old man warmly clothed and wrapped in his mantle.

European Centre for Medium-Range Weather Forecasts (ECMWF)—The forecast center located in Reading, England, and formed by a group of European countries as a joint venture to carry out research into improving numerical weather forecasting for up to 10 days in advance and to produce such **medium-range forecasts** on an operational basis.

eutrophication—The enrichment of water by nutrients, especially compounds of nitrogen and phosphorous, that will accelerate the growth of algae and other higher plant life forms. A deficiency of dissolved oxygen results from this natural or anthropogenically accelerated process.

evaporation—(1) The physical process by which a liquid is transformed to the gaseous state; the opposite of **condensation**. **Latent heat** of evaporation is consumed in the process, being stored in the water vapor, with a cooling of the surroundings. In meteorology, evaporation usually is restricted in use to the change of water from liquid to gas at temperatures below the **boiling point**, while **sublimation** is used for the change from solid to gas. (2) The amount of water evaporated per unit area, often described in depth units (e.g., inches or millimeters).

evaporation pan—An **evaporimeter**, or large pan used in the measurement of the **evaporation** of water into the atmosphere during a given time interval; typically, these are located at agricultural stations. The National Weather Service evaporation pan (*class-A pan*) is a cylindrical container fabricated of galvanized iron or monel metal with a depth of 10 inches (25.4 centimeters) and a diameter of 48 inches (121.9 centimeters).

evaporimeter—An instrument used for measuring the quantity of water substance evaporated into the atmosphere during a given time interval. See **evaporation pan**.

evapotranspiration—The **vaporization** of water through direct **evaporation** from wet surfaces plus the release of water vapor by plants through leaf pores (**transpiration**). See **actual** and **potential evapotranspiration**.

exceedance interval—The average number of years between the occurrence of an event and that of a ''greater'' event. See **return interval**.

excessive heat warning—Issued by the National Weather Service to warn the public that **heat indices** are expected to be extremely high, typically 115°–120°F (46°–49° C) or higher for at least 3 hours.

excessive heat watch—Issued by the National Weather Service to inform the public that high temperatures and high humidities leading to high **heat indices** are possible within the next several days.

excessive precipitation—Any **precipitation** or **heavy rain** event that falls at an unusually high accumulation rate that exceeds limits based upon time of duration.

exhalation—The process by which radioactive gases escape from the surface layers of soil or loose rock where they are formed by decay of radioactive materials.

exhaust trail—A **condensation trail** (**contrail**) that forms when the water vapor of the aircraft exhaust is mixed with and saturates (or slightly supersaturates) the air in the wake of the aircraft.

exogenic influences—Influences on the earth's **climate** that originate outside its **atmosphere** or **hydrosphere**, (e.g., solar or cosmic radiation, comets, etc.).

exosphere—The uppermost layer of the earth's **outer atmosphere**, its lower boundary is estimated at 500–1000 kilometers above the earth's surface. It is only from the exosphere that atmospheric gases can, to any appreciable extent, escape into outer space.

exothermic—A process involving the release of heat **energy**. The reverse of **endothermic**.

exposure—The general surroundings of a weather observation site, instruments, or other apparatus, with special reference to its openness to winds and sunshine. An important consideration in the representativeness of observed data.

exsiccation—Drying up by the removal of moisture. In climatology it implies the loss of moisture by draining or by increased **evaporation**, rather than by a decreased supply of water from precipitation (**desiccation**).

extended forecast—In general, a forecast of weather conditions for a period extending beyond two days from the day of issue. Same as **medium-range forecast** or **long-range forecast**.

extinction—In general, the termination. The **attenuation** of light; that is, the reduction in **illuminance** of a parallel beam of light as the light passes through a medium wherein **absorption** and **scattering** occur.

extraterrestrial radiation (or *extra-atmospheric radiation*)—In general, **solar radiation** received at the limit of the earth's atmosphere.

extratropical—In meteorology, weather occurrences poleward of the belt of **tropical easterlies**.

extratropical cyclone—Any **synoptic-scale** storm system that is not a **tropical cyclone**, usually referring only to the migratory **frontal cyclones** of middle and high latitudes.

extratropical low—Same as **extratropical cyclone**.

extratropical storm—Same as **extratropical cyclone**.

extreme—In climatology, the highest and, in some cases, the lowest value of a **climatic element** observed during a given time interval or during a given month or season of that period. If this value were the most extreme for the whole **period of record** for which observations are available, the value would be the *absolute extreme*.

eye—In meteorology, usually the ''eye of the storm'' (**hurricane**, **typhoon**); that is, the roughly circular area of comparatively light winds and fair weather found at the center of a severe **tropical cyclone**. The eye is surrounded by an **eyewall**. The winds in the eye are generally 10 knots or less; no rain occurs; sometimes blue sky may be seen because of **subsidence** (sinking) of air within the eye. Eye diameters typically vary from 20 to 65 kilometers.

eyewall—The organized ring of intense thunderstorms surrounding the **eye** of a **tropical cyclone**, typically a **hurricane**. This region experiences the most severe weather with heavy precipitation and high winds. Sometimes the eyewall is also called a **wall cloud**, but use of the latter term in this present context is not encouraged.

F

F-scale—Abbreviation for **Fujita scale**.

F_1-layer—The "layer" of the **ionosphere** that exists as an appendage on the lower part of the **F_2-layer** during the day, approximately 200–300 kilometers in altitude. It exhibits a distinct maximum of **free-electron** density except at high latitudes in winter, when it is not detectable.

F_2-layer—The highest permanently observable "layer" of the **ionosphere**. It exhibits a distinct maximum of **free-electron** density occurring at an altitude of about 225 kilometers in the polar winter to over 400 kilometers in the daytime near the **magnetic equator**.

faculae—Large patches of bright material appearing on the sun's **photosphere** forming a veined network in the vicinity of **sunspots**. They appear to be more permanent than sunspots and are probably due to elevated clouds of luminous gas.

Fahrenheit temperature scale—A temperature scale, primarily used by the public in the United States, on which water freezes at 32° and water boils at 212° (at sea level air pressure). Named for Daniel Fahrenheit (1686–1736), a German physicist who devised this scale.

fair—With respect to **weather**, generally descriptive of pleasant weather conditions, with due regard for location and time of year. It is subject to popular misinterpretation, for it is a purely subjective description. When this term is used in public weather forecasts (National Weather Service), it is meant to imply (*a*) no precipitation, (*b*) less than 0.4 **sky cover** of low clouds, and (*c*) no other extreme conditions of cloudiness, visibility, or wind.

fair-weather cumulus—Same as **cumulus humilis** clouds.

Falkland Current—In the South Atlantic Ocean, a cold ocean current flowing northward along the Argentine coast. The Falkland current originates as a branch of **the Antarctic Circumpolar Current**, in the vicinity of the Falkland Islands. At about 35°S it is joined by the **Brazil Current**, both flowing east across the ocean as the **South Atlantic Current**.

fall speed—Normally the speed at which precipitation elements are falling through the adjacent air, up to 9 meters per second for the largest **raindrops** and much greater for **hail**. See **terminal velocity**.

fall wind—A cold and usually downslope wind, which differs from a **foehn** (or **Chinook**) in that the air initially was sufficiently cold to remain relatively cold at surface despite compressional heating during descent. It differs from the smaller-scale

gravity wind in that a cold pool of air typically accumulates at high elevations. Examples of fall winds include the **bora** and the **mistral**.

falling limb—The portion of the **hydrograph** indicating a decrease in **discharge**. Contrast with **rising limb**.

falling tide (also known as *ebb tide*)—The portion of the **tide cycle** between **high water** and the following **low water**, marked by a fall in water level. Contrast with **rising tide**.

fallout—The descent to the ground of dust and other debris raised to great heights in the atmosphere by a violent explosion, especially applied to radioactive fallout from an atomic or thermonuclear explosion.

fallstreaks—Same as **virga**.

false echo—A **radar echo** on a radar screen that does not represent a normal hydrometeor target, such as rain or snow, but may be associated with strong, nonstandard variations in atmospheric **refractive index,** birds, insects, etc. See **anomalous propagation**.

fanning—A pollution **plume** that is carried downwind with lateral **dispersion** exceeding vertical dispersion from the downwind axis, forming a fan-shaped appearance because of the deep stable atmospheric layer extending above the plume. Compare with **coning**.

far infrared—Pertaining to the region of the spectrum with **electromagnetic radiation** consisting of wavelengths greater than 4 micrometers, or beyond the **middle infrared**, but shorter than **microwave** radiation (0.1 millimeters or longer), and **radio energy**.

fast ice—Any fixed type of sea, river, or lake ice attached to the shore (**ice shelf**), **beached (shore ice**), stranded in shallow water, or frozen to the bottom of shallow waters (**anchor ice**).

Fata Morgana—An impressive and complex **mirage** with multiple distortions of images, especially in the vertical. Originally named for the optical phenomena seen over the Straits of Messina where cold air over warm water gave the appearance of fairy castles on the far shores.

fathom—The traditional unit of depth in the ocean, equal to six feet.

feather—See **barb**.

feedback—Atmospheric processes where a change in a variable, through interactions in the system, either reinforces the original process (*positive feedback*) or suppresses the process (*negative feedback*).

Ferrel cell—A shallow atmospheric circulation cell in the middle and high latitudes of both hemispheres, proposed in 1859 by William Ferrel (1817–1891), an American meteorologist, in which the flow is poleward near the ground and equatorward at an intermediate level; not observable on mean charts.

fetch—(1) The area in which ocean waves are generated by the wind. It is generally delineated by coastlines, **fronts**, or areas of wind curvature or **divergence**. (2) Distance measured along a large water surface **trajectory** over which a wind of almost uniform direction and speed blows—that is, in the direction of the wind.

field—A region of space at each point of which a given physical or mathematical quantity has some definite value. Thus, one may speak of a gravitational field, **magnetic field**, or **electric field**; and, in meteorology, we speak of a pressure field, temperature field, and wind field.

field capacity—Amount of water held in the soil after **gravitational water** has drained away.

filling—A rise in the **central pressure** of an atmospheric pressure system on a **constant-height chart**, or an analogous increase in height (specifically, **geopotential height**) on a **constant-pressure chart**. Opposite of **deepening**. The term is commonly applied to a low rather than to a high.

fire weather—A combination of weather conditions that favors the kindling and spread of forest or brush fires; typically, low **humidity** and a lack of recent precipitation.

fireball—A very bright **meteor**; also, the luminous cloud of dust and vapor accompanying a nuclear explosion.

firn—Old snow that has become granular and compacted (dense) as the result of various surface metamorphoses, mainly melting and refreezing but also including **sublimation** and **deposition**. The resulting particles are generally spherical and rather uniform. Ultimately, firn is often transformed into **glacier ice**. Some authorities restrict the use of *firn* to snow that has lasted through one summer, thereby distinguishing it from **spring snow**.

firn line—The boundary on a **glacier's** surface between its **accumulation zone** (above) and **ablation zone** (below).

firnification—The process of **firn** formation; the first step in the transformation of snow into **glacier ice**.

first law of thermodynamics—A statement of the **conservation of energy** for thermodynamic systems (not necessarily in equilibrium). The fundamental form requires that the heat absorbed by the system serves either to raise the **internal energy** of the system or to do **work** on the environment.

first-order climatological station—As defined by the **World Meteorological Organization**, a meteorological station at which autographic records or hourly readings of atmospheric pressure, temperature, humidity, wind, sunshine, and precipitation are made, together with observations at fixed hours of the amount and form of clouds and notes on the weather.

first-order station—After National Weather Service practice, any meteorological station that is staffed in whole or in part by National Weather Service personnel, regardless of the type or extent of work required of that station.

fixed balloon—A tethered balloon carrying instruments for making observations in the lower layers of the atmosphere; it is connected by a cable to a raising and lowering apparatus.

flare—See **solar flare**.

flash flood—A **flood** (1) that rises and falls quite rapidly with little or no advance warning, usually as the result of intense rainfall over a relatively small area; other possible causes are **ice jams**, dam failure, etc. Usually it occurs within 6 hours of a rain event and it has a high **peak discharge**. Due to the hazards, the National Weather Service issues **flash flood watches** and **warnings** to the public.

flash flood warning—Issued by the National Weather Service to warn the public that a dangerous **flash flood** situation is imminent in the affected area or is occurring due to heavy rain or dam failure. The public is urged to take appropriate action immediately. Contrast with **flash flood watch.**

flash flood watch—Issued by the National Weather Service to alert the public that **flash floods** are a possibility in or close to the designated watch area. The public in the affected area is urged to be ready to take action in the event that a **flash flood warning** was issued or flooding was observed. Contrast with **flash flood warning** and **urban and small stream flood advisory.**

floe—A chunk of ice floating on the surface of a water body; may consist of a single fragment or many fragments.

floeberg—A mass of ice formed by the piling up of many ice **floes** as a consequence of lateral pressure; an extreme form of **pressure ice**; it may be more than 15 meters high and resemble an **iceberg**.

flood—(1) The condition that occurs when water overflows the natural or artificial confines of a stream or other body of water, or accumulates by drainage over low-lying areas. (2) Any controlled increase in water that spreads across an area. (3) Rising tide; see **flood tide**.

flood current—The movement of a **tidal current** toward the coast, or up an **estuary** or tidal waterway; erroneously called **flood tide**. Opposite of **ebb current**.

flood forecast—Prediction of the **stage**, **discharge**, beginning, and duration of a **flood**, especially of the **peak discharge** at a specific point on a stream resulting from precipitation and/or snowmelt. See also **urban and small stream flood advisory**, **flash flood watch**, and **flash flood warning**.

flood plain—Any part of a valley floor subject to occasional **floods** that may threaten life and property.

flood routing—Process of determining progressively the timing and shape of a **flood wave** at successive points along a river or throughout a reservoir.

flood stage—That level (**stage**), on a fixed **river gauge**, at which overflow of the natural banks of a stream begins to cause damage in any portion of the **reach** for which the gauge is used as an index.

flood tide—(1) Also called **rising tide**; the part of the **tide cycle** between **low water** and the following **high water**. (2) A tide at its highest point.

flood wave—A rise in streamflow in a river channel to a maximum crest at **flood stage**, followed by recession.

Florida Current—As part of the **Gulf Stream system** of the western North Atlantic, the Florida Current consists of all northward-moving warm water from the Straits of Florida to a point off Cape Hatteras, North Carolina, where the current ceases to follow the **continental slope**. It is one of the swiftest of ocean currents (flowing at a rate of 2–5 knots). The Florida Current can be traced directly back to the Yucatan Channel because the water flowing through the channel continues on the shortest route to the Straits of Florida and only a small amount sweeps into the Gulf of Mexico, later to join the Florida Current. After passing the Straits of Florida the current is reinforced by the **Antilles Current**, but the name Florida Current is retained as far as Cape Hatteras.

flow pattern—Distribution of velocities in a region of the atmosphere at a particular time.

fluid—Any matter that flows or conforms to the confines of a container; that is, a **gas** or a **liquid**.

flurry—(1) A popular term for sudden **snowshowers**, usually with little accumulation. (2) A sudden and brief wind **squall**.

flux—The time rate of flow of some quantity, often used in reference to the flow of some form of **energy**, especially **radiation**. See **transport**.

foehn (or **föhn**)—A European term for a warm, dry wind that is drawn down the Alpine valleys of Austria and Germany. The warmth and dryness of the air is due to adiabatic compression upon descending the mountain slopes. Often times used as the generic term for such a downslope warm wind. See **Chinook.**

fog—A visible aggregate of minute water droplets suspended in the atmosphere near the earth's surface; a **cloud** in contact with the earth's surface. Fog is responsible for reducing visibility to less than 1 kilometer (5/8 mile). Contrast with **haze**, **mist**, and **ground fog**. Several types of fogs can be classified according to formation, to include **advection fog**, **frontal fog**, **upslope fog**, and **radiation fog**.

fog advisory—Issued by the National Weather Service to make the public aware that dense fog covers a widespread area and reduces the visibility to 0.25 mile (400 meters) or less; may also be called a **dense fog advisory**.

fog bank—Generally, a fairly well-defined mass of **fog** observed in the distance, most commonly at sea. This term is not applied to ''patches'' of shallow fog.

fog drip—Water dripping to the ground from trees or other objects, which have collected the moisture from drifting **fog**. Makes a significant contribution to the maintenance of California's near-coastal forests.

fogbow—An optical phenomenon that is essentially a **primary rainbow** consisting of a luminous white band appearing on a fog screen, centered on the observer's **antisolar point**. The band may be fringed with red on the outside and blue on the inside.

force—The agent that changes the state of rest or motion of a mass when applied; the time rate of change of **momentum** of a body of unit mass. The **SI** unit of force is the **newton**.

forced convection—Mass motions within a **fluid** resulting in the **transport** and **mixing** of the properties of that fluid and induced by mechanical forces such as deflection by a large-scale surface irregularity. Contrast with **free convection**.

forecast—A definite statement or statistical estimate of expected future occurrences; that which is predicted.

forecast lead time—The time interval between the issuance of the forecast and the start of the scheduled **forecast period** or, in the case of a **warning**, the occurrence of the anticipated event.

forecast period—The time interval for which a **forecast** is made; see **verification time**.

forecast verification—Any process for determining the accuracy of a weather **forecast** by comparing the predicted weather with the observed weather of the forecast period. Principal purposes of forecast verification are to test forecasting skills and methods.

forerunner—Low, long-period, ocean **swell** that commonly precedes the main swell from a distant storm, especially a **tropical cyclone**.

forest wind—A light breeze that blows from forests toward open country on calm clear nights. The principal **nocturnal cooling** in a forest takes place at the leaves and branches of the tree tops. This cold forest air sinks through the trees and then flows laterally toward open ground. The return current takes places at a level slightly above the tree crowns.

forked lightning—A common form of **lightning discharge**, in a **cloud-to-ground discharge**, that exhibits downward-directed branches from the main **lightning channel**.

forward scatter—The **scattering** of **radiant energy** into the hemisphere of space bounded by a plane normal to the direction of the incident radiation and lying on the side toward which the incident ray is traveling. Opposite of **backscatter**. The size of the scattering agent relative to the wavelength of incident radiation affects the ratio of forward to total scatter; half the scattered radiation is forescattered in **Rayleigh** scatter involving relatively small scattering agents, while the ratio increases as the size of the **aerosol** relative to radiation wavelength increases, especially in **Mie scatter**.

fossil ice—Ice found in regions of **permafrost**, or in other regions where present-day temperatures are not low enough to have formed it; ice that was formed in the geologic past.

fragmentation—The process of breakup of an **ice crystal** into a large number of fragments, each of which can serve as a new center of growth (*fragmentation nucleus*) for other crystals.

frazil—Fine, needlelike or plate-shaped ice crystals that form and are suspended in supercooled water that is too turbulent to permit coagulation into **sheet ice**. This is most common in swiftly flowing streams, but also is found in a turbulent sea. It may accumulate as **anchor ice** on submerged objects obstructing the water flow.

free atmosphere—That portion of the earth's atmosphere, above the **planetary boundary layer** (1–2 kilometers), in which the effect of the earth's **surface friction** on the air motion is negligible, and in which the air is usually treated (dynamically) as an ideal **fluid**.

free convection—Mass motions within a **fluid** resulting in the **transport** and **mixing** of the properties of that fluid and caused only by density differences within the fluid. Contrast with **forced convection**.

free electron—An electron or negative electric charge that is not bound to an **ion**; found in the **ionosphere**, where the extrema in density help determine the various layers.

free radical—Very reactive atoms and or molecules with unpaired electrons. Usually these free radicals are short-lived intermediate species in the atmosphere, produced by **photodissociation** of source molecules by solar **ultraviolet** radiation or by reactions with other stratospheric constituents. Many stratospheric chain reactions involving the destruction of ozone molecules have free radicals as intermediate species.

freeze—The condition that exists when, over a widespread area, the surface temperature of the air remains below freezing 0°C (32°F) for a sufficient time to constitute the characteristic feature of the weather. A freeze is a term used for the condition when vegetation is injured by these low air temperatures, regardless if **frost** were deposited. Freezes may be classified as *light freeze* (little destructive damage, except to tender plants); *killing freeze* (widely destructive to vegetation, effectively terminating the **growing season**); and *hard freeze* (staple vegetation destroyed).

freeze warning—Issued by the National Weather Service during the **growing season** to make agricultural interests and the public aware of anticipated **freeze** conditions since the temperatures are expected to drop well below 0°C (32°F) over a large area, whether or not **frost** forms. Compare with **frost advisory**.

freeze-up—Formation of **ice cover** on a large water surface (e.g., lake) due to seasonal cooling in late autumn or winter. A phenological event, recorded in northern United States and Canada.

freezing—(1) The phase transition of a substance passing from the liquid to the solid state; solidification. In meteorology, this almost invariably applies to freezing water. **Latent heat** of freezing is released by the water molecules and heats the system. (2) The environment when the **ambient temperature** is equal to or less than 0°C (32°F).

freezing drizzle—A **freezing precipitation** form where **drizzle** that falls in liquid form becomes supercooled and freezes upon impact with cold surfaces to form a coating of **glaze**. See **freezing rain**.

freezing fog—A **fog** whose supercooled droplets freeze upon contact with exposed cold objects and form a coating of **rime** and/or **glaze**.

freezing level—Commonly, and in aviation terminology, the lowest altitude in the atmosphere, over a given location, at which the air temperature is 0°C (32°F); the altitude of the 0°C (32°F) constant-temperature surface. Essentially the same as **melting level**, a term used in cloud physics.

freezing-level chart—A synoptic chart showing the lowest altitude of the 0°C (32°F) constant-temperature surface by means of **contour** lines.

freezing nucleus—Any particle that, when present within a mass of supercooled water, will initiate growth of an ice crystal about itself (see **nucleation**). Contrast with **cloud condensation nuclei**.

freezing point—The temperature at which a liquid solidifies under any given set of conditions; this temperature may not necessarily be identical to that of the **melting point**. See **ice point**.

freezing precipitation—Any form of **liquid precipitation** that freezes upon impact with the cold ground or exposed object surfaces at subfreezing temperatures (0°C or 32°F or below)—that is, **freezing rain** or **freezing drizzle**. Contrast with **frozen precipitation**.

freezing rain—A **freezing precipitation** form where **rain** that falls in liquid form becomes supercooled and freezes upon impact with cold surfaces (at subfreezing temperatures) to form a coating of **glaze** upon the ground and on exposed objects; this type of precipitation is responsible for **ice storms**. See **freezing drizzle**.

freezing rain (or **freezing drizzle**) **advisory**—Issued by the National Weather Service to alert the public of the potential for **freezing rain** or **freezing drizzle** of sufficiently light intensity and that ice is not expected to form on all exposed surfaces. Contrast with **ice storm warning**.

frequency—The rate of recurrence of an event in periodic motion. Frequency has the dimension of reciprocal time and is usually expressed according to convenience in units of radians per second, oscillations (cycles) per second, or **Hertz**; the reciprocal of the **period**.

fresh breeze—In the **Beaufort wind scale** (*Beaufort force* 5), a wind whose speed is 17–21 knots (19–24 mph), causing small leafy trees to sway and moderate waves to form.

fresh gale—Former name for **gale** (*Beaufort force* 8).

freshwater—Naturally occurring water having a low concentration of dissolved salts; generally acceptable for use as potable water. In oceanography, the differences in density between freshwater and **saltwater** or **brine** produces an oceanic frontal zone.

friction—The mechanical resistive **force** offered by one medium or body to the relative motion of another medium or body in contact with the first.

friction layer—Same as **planetary boundary layer**.

Frigid Zone—Based on ancient **climatic classification** scheme using solar illumination geometry, the region poleward of the **Arctic** or **Antarctic Circles** (66° 33' North and 66° 33' South, respectively). Contrast with **Temperate** and **Torrid Zones**.

front—In meteorology, generally, the **interface** or transition zone between two **air masses** of different density. A front usually separates air masses of different temperature and moisture. Since the temperature distribution is the most important regulator of atmospheric density, fronts usually are found along regions of strong temperature contrast. More specifically, a front represents the intersection of the transition zone with the earth's surface. Fronts can be classified on a **synoptic surface chart** according to the direction the front is moving with respect to an air mass: **cold front**, **warm front**, **stationary front**, or **occluded front**. Fronts also can be identified as **arctic** or **polar front**, according to the adjoining air mass.

frontal cyclone—In general, any organized atmospheric low pressure system (**cyclone**) associated with a **front**; often used synonymously with **wave cyclone** or with **extratropical cyclone** (as opposed to **tropical cyclones**, which are nonfrontal).

frontal fog—A **fog** type associated with **frontal zones** and **frontal passages**.

frontal inversion—A **temperature inversion** in the atmosphere, encountered upon vertical ascent through a sloping **front** (or *frontal zone*).

frontal lifting—The forced ascent of the warmer, less-dense air at and near a **frontal surface**, occurring whenever the relative velocities of the two **air masses** are such that they converge at the front; typically associated with **warm fronts**.

frontal passage—Movement of a **front** past a specific place. It is often accompanied by characteristic sequence of changes in **temperature**, **pressure tendency**, **wind direction**, and **speed**, **visibility**, **cloudiness**, and **precipitation**.

frontal precipitation—Any precipitation attributable to the action of a **front**; used mainly to distinguish this type from **orographic precipitation** and **airmass showers**.

frontal profile—The outline of a **front** as seen on an analyzed vertical **cross section** oriented perpendicular to the frontal surface.

frontal surface—See **front**.

frontal system—An assemblage of **fronts** as they appear on a **synoptic chart**. This is used for (*a*) a continuous front and its characteristics along its entire extent, including its **warm**, **cold**, **stationary**, and **occluded** sectors, its variations of intensity, and any frontal cyclones along it; and (*b*) the orientation and nature of the fronts within the circulation of a **frontal cyclone**.

frontal theory—A conceptual model of the formation and development of **air masses** and **fronts** in the atmosphere, and the formation and development of extratropical **depressions** in relation to air masses and fronts.

frontal thunderstorm—A thunderstorm associated with a **front**. Strictly, this term should be limited to thunderstorms resulting only from the convection induced by **frontal lifting**.

frontal zone—See **front**.

frontogenesis—The initial formation of a **front** or **frontal zone**.

frontolysis—The dissipation of a **front** or **frontal zone**.

frost—(1) A cover of ice crystals produced by **deposition** of atmospheric water vapor directly upon a surface at or below 0°C. Ice crystals may be of many forms. Not to be confused with frozen **dew** (**white dew**) or **rime**. See **hoarfrost**. (2) The condition that exists when the temperature of the earth's surface and earth-bound objects falls below freezing, 0°C or 32°F. Depending upon the actual values of ambient-air temperature, **dewpoint**, and the temperature attained by surface objects, frost may occur in a variety of forms. These include a general **freeze**, hoarfrost (or **white frost**), and **black frost**. If a frost period were sufficiently severe to end the **growing season** (or delay its beginning), it is commonly referred to as a *killing frost*.

frost action—In general, cycles of freezing and thawing of water contained in natural or artificial materials. This is especially applied to the disruptive effects of this action.

frost advisory—Issued by the National Weather Service during the **growing season** to make agricultural interests and the public aware that the formation of widespread **frost** is anticipated. Compare with **freeze warning**. A frost warning may be issued for a killing frost.

frostbite—The freezing of exposed parts of the body, especially the extremities, causing local damage to the tissues.

frostburn—Damage to skin tissue resulting from contact of bare skin with metal surfaces at below-freezing temperatures.

frost damage—Damage to vegetation occurring when the water that is part of the cell structure of the plant solidifies, bursting cell walls and deteriorating the plant materials; occurs with a *killing frost* (see discussion in **frost**).

frost hazard—The risk of damage by **frost**. It may be expressed as the probability or frequency of killing frost on different dates during the **growing season**, or as the distribution of dates of the last *killing frost* of spring or the first of autumn.

frost heaving—The lifting of a surface by internal **frost action**, the extreme result of which is a **frost mound**. Most frost heaving has been found to result from the growth of lenticular masses of ice within the soil.

frost mound—A conical mound on a land surface, caused by the freezing of water in the ground. It is a product of **frost heaving**, but is unusual in that a great concentration of water is required in a relatively small subsurface volume. Usually a frost mound is of seasonal duration, while a **pingo** is more than one year duration.

frost pocket (or *frost hollow*)—A topographic depression with sufficiently different vegetation to indicate a **microclimate** that experiences a greater frequency of **frost** or **freeze** events, resulting in a markedly shorter **growing season** than the surrounding terrain. These depressions collect cold dense air draining from higher terrain.

frost point—The temperature to which air must be cooled at constant pressure and constant humidity to achieve **saturation** with respect to ice at or below 0°C (32°F). See **dewpoint**.

frost-point hygrometer—An instrument for measuring the **frost point** of the atmosphere. Air is passed continuously across a polished surface whose temperature is adjusted so that a thin deposit of frost is formed that is in **equilibrium** with the air. The surface temperature is varied by directing a jet of cooled liquid against the under surface or by passing an electrical current through a heating coil surrounding the surface; measurement of this temperature usually is made with a **resistance thermometer**.

frost protection—Strategies for preventing damage or loss of fruit, trees, or plants due to subfreezing temperatures. Examples are flooding cranberry bogs, radiation screens to trap heat, fans for **mixing** air, and water spray on citrus trees.

frost table—An irregular surface in the ground that, at any given time, represents the penetration of thawing into seasonally frozen ground. In regions of **permafrost**, when the **active layer** is thawed completely this coincides with the **permafrost table**.

frost zone—The layer of ground subject to seasonal freezing. In regions of **permafrost**, this corresponds to the **active layer**.

frozen precipitation—Any form of precipitation that reaches the ground in frozen form (e.g., **snow**, **snow pellets**, **snow grains**, **ice crystals**, **ice pellets**, **hail**). Contrast with **freezing precipitation**.

Fujita scale (F-scale)—Tornado intensity scale developed by T. Theodore Fujita of the University of Chicago; classifies **tornadoes** from F0 (gale tornado with wind speeds 40–72 mph) to F5 (incredible tornado with wind speeds 261–318 mph) on the basis of rotational wind speeds estimated from the damage they caused.

fulgurite—A glassy, rootlike tube formed when a **lightning stroke** terminates in dry sandy soil. The intense heating of the current passing down into the soil along an irregular path fuses the sands. Concurrently, vaporization of **soil moisture** and possibly even vaporization of the sandy materials, causes the molten material to be expanded into a tube whose diameter may be over an inch, but whose wall is very thin. Fulgurites have been recovered in lengths of over five feet.

fumigation—A process by which air pollutants accumulated beneath a stable upper-air layer are transported down to the ground by **convective mixing**. Usually this situation occurs with unstable conditions in the lower atmosphere below the stack level and stable conditions aloft. Compare with **trapping**.

fundamental unit—A unit measure of a basic physical quantity such as mass, length, time, temperature—for example, one **kilogram**, one **meter**, one **second**, one **kelvin**, respectively. Other quantities such as **force**, **speed**, etc., may be considered *derived units*, formed by combining fundamental units according to appropriate algebraic relationships; these derived units may have approved special names (e.g., **newton** for force and **pascal** for pressure).

funnel cloud—A rapidly rotating column of air (a tornadic circulation) extending below cloud base but not reaching the ground; made visible by condensation to form a cone-shaped cloud; a funnel cloud would be classified a **tornado** if it touched the ground.

fusion—(1) A process involving **melting**, where a physical phase change occurs between the solid and liquid phases of a substance. Heat or pressure are required. (2) A nuclear process where two or more nuclei combine at high temperatures to form a heavier nucleus with the release of energy.

G

gale—(1) In general, and in popular use, an unusually strong wind. (2) In the modern version of the **Beaufort wind scale** (*Beaufort force* 8), a wind whose speed is from 34 to 40 knots (39 to 46 mph), twigs break from trees and moderately high waves are produced.

gale warning—A storm **warning**, for marine interests, of impending winds from 34 to 47 knots (39 to 54 mph). Typically used for conditions not directly associated with **tropical cyclones**. The storm warning signals for this condition are (*a*) two triangular red pennants by day, and (*b*) a white lantern over a red lantern by night.

gamma rays—**Electromagnetic radiation** of very short wavelength (0.001–0.1 nanometers) and extremely high energy level; emitted by atomic nuclei in the course of certain radioactive disintegrations. Compare **alpha particle** and **beta particle**.

gas—One of the three physical **phases** of matter; a gas is a **fluid** containing molecules that are sufficiently mobile to expand indefinitely to completely fill the container. Compare with **liquid** and **solid**.

gas constant—The proportionality constant in the **equation of state** for **ideal gases**. It can be specified as the *universal gas constant*, independent of the gas species, such that for one **mole** of an ideal gas, the value is $R^* = 8.31433$ joules per kilomole per kelvin. It can also be specified as the *specific gas constant* $R=R^*/M$ for a unit mass of a particular gas species of given molecular weight M, the equation is $pV=RT$, where p is the pressure; V, the volume; R, the specific gas constant; and T, the temperature (absolute). For dry air, the value of R is 287.054 joules per kilogram per kelvin.

gas laws—The thermodynamic laws applying to **perfect gases**: **Boyle's law**, **Charles's law**, **Dalton's law**, **equation of state**.

gauge relation—An empirical curve relating stream **discharge** or **stage** at a point on a stream to discharge or stage at one or more upstream points and, possibly, to other parameters.

general circulation (also known as *global circulation*)—In its broadest sense, the complete statistical description of atmospheric motions over the earth. These statistics are generated from the ensemble of daily flow patterns, and include not only the temporal and spatial *mean* conditions (e.g., zone westerlies and easterlies, semipermanent waves, and meridional cells), which are sometimes called the general circulation, but also the higher-order statistics, which measure the spatial and temporal *variability* of the flow resulting from seasonal changes and from the effects of transient cyclones and anticyclones. See discussion in **planetary circulation**.

general circulation model (**GCM**)—Numerical representation of the atmosphere and its phenomena over the entire earth, using the **equations of motion** and including radiation, photochemistry, and the transfer of heat, water vapor, and momentum.

gentle breeze—In the **Beaufort wind scale** (*Beaufort force* 3), a wind whose speed is from 7 to 10 knots (8 to 12 mph), causing movement of leaves and scattered **whitecaps**.

geologic age—A formal **geologic time** unit representing a subdivision of a **geologic epoch**, corresponding to a **stage**.

geologic epoch—A **geologic time** unit with a time interval longer than a geologic age, but shorter than a geologic period. The **Holocene epoch** is the most recent epoch in the **Quaternary period**.

geologic era—The longest time unit in **geologic time**. The Cenozoic era is the most recent era, spanning the last 65 million years.

geologic period—A major worldwide **geologic time** unit; a subdivision of a **geologic era**, but longer than a geologic epoch. The **Quaternary Period** contains the most recent period in the Cenozoic era containing the **Pleistocene** and **Holocene epochs**.

geologic time—(1) Popularly, a span of millions or billions of years in the past, prior to the start of modern history. (2) Any formal division or unit of geological chronology; for example, **geologic era**, **period**, **epoch**, or **age**.

geomagnetic pole—One of the two points on the earth's surface representing a pole of the system of geomagnetic coordinates; that is, an axis pole of the mathematical **magnetic field** of closest fit to the actual magnetic field of the earth; this pole tends to coincide with the center of the **auroral oval**. At present, the north geomagnetic pole is near Thule, Greenland (79°N, 71°W), while the south geomagnetic pole is near Vostok, Antarctica (79°S, 109°E). Compare with the **magnetic pole**.

geomagnetism—The magnetic phenomena, collectively considered, exhibited by the earth and its atmosphere.

geomorphology—The explanatory science dealing with the form and surface configuration of the solid earth. It is primarily an attempt to reveal the complex interrelationships between the origin (and therefore material composition) of surface features on the one hand, and the causes of the surface alteration (**erosion**, **weathering**, crustal upheaval, etc.) on the other hand.

geophysics—The interdisciplinary study of the physics of the earth and its environment—that is, earth, air, and (by extension) space.

geopotential—The **potential energy** of a unit mass relative to sea level, numerically equal to the **work** that would be done in lifting the unit mass from sea level to the altitude at which the mass is located. See **geopotential height**.

geopotential height—A measure of the **altitude** of a point in the atmosphere, expressed in terms of its **potential energy** per unit mass (**geopotential**) at this altitude relative to sea level. Units are geopotential meters. Geopotential height is used in essentially all operational **upper-air** reports and charts. Differences between geopotential and traditional geometric height are minor (less than 0.5%) for the lowest 30 kilometers in the atmosphere.

Geostationary Operational Environmental Satellite (GOES)—A **geosynchronous satellite** series designed to monitor weather systems. GOES observes the United States and adjacent ocean areas from a vantage point of 35 790 kilometers (22 240 miles) above the earth's surface along the equator. Digital data from onboard **scanning radiometers** made in the visible, infrared, and water vapor channels are sent back to earth stations from this satellite for the production of **satellite images**. Vertical temperature **soundings** are also made using **sounders**.

geostrophic current—In oceanography, a hypothetical ocean current associated with horizontal **pressure gradients** in the ocean and determined by the condition that the **pressure force**, due to the distribution of mass, balances the **Coriolis effect** due to the earth's rotation. The geostrophic current corresponds to the **geostrophic wind** in meteorology.

geostrophic wind—A hypothetical model describing an unaccelerated horizontal wind that blows in a straight path parallel to **isobars** or **contours** above the **friction layer**; it results from a balance between the horizontal components of the **pressure gradient force** and the **Coriolis effect**. The speed is inversely proportional to isobar (contour) spacing. In the Northern Hemisphere, the geostrophic wind is oriented such that lower pressure is to the left of the wind.

geosynchronous satellite (also known as *geostationary satellite*)—A satellite that orbits earth at the same rate and in the same direction as the earth rotates so that the satellite is always directly over the same point (the **subsatellite point**) on the earth's equator. The altitude of the satellite's orbit is about 35 790 kilometers (22 240 miles). From this position, approximately one-third of the earth can be effectively imaged. Contrast with **polar-orbiting satellite**.

glacial maximum—The time and/or position of the greatest extent of any continental-scale **glacier**. It is most frequently applied to the greatest equatorward advance of **Pleistocene glaciation**.

glacial stage—A major subdivision of the **Pleistocene glaciation**.

glaciation—(1) Alteration of any part of the earth's surface by passage of a **glacier**, chiefly by glacial erosion or deposition. (2) An interval of time with extensive **continental glaciers**—for example, the **Pleistocene glaciation**. (3) The transformation of cloud particles from water drops to ice crystals. Thus, a **cumulonimbus** cloud is said to have a ''glaciated'' upper portion (especially in the **anvil**).

glacier—A mass of freshwater ice of atmospheric origin that is formed on land as a result of the further recrystallization of **firn**, flowing slowly in response to **gravity** (at present or in the past) from an **accumulation zone** to an **ablation zone**. While its origin is on land, a glacier may terminate over water bodies.

glacier flow—The slow motion of elements of a **glacier** downward and outward because of gravitational forces; changes in the areal size of glacier are not necessary.

glacier ice—Any ice that is or was once a part of a **glacier**. It has been consolidated from **firn** by further melting and refreezing, and by **static pressure**. Glacier ice may be found in the sea as an **iceberg**.

glacier wind—A shallow **gravity wind** that blows downslope along the surface of the **glacier**; the wind, which may intensify in the afternoon, is a result of the temperature contrasts between the air in proximity to the glacier slope and the free atmosphere at the same altitude.

glaciology—The study of the properties and occurrence of snow and ice on the earth's surface, with specific concentration on the regime of active **glaciers**.

glare—Any hindrance to vision caused by **scattering** or **reflection** of light into an observer's line of sight.

glaze—A coating of ice, generally clear and smooth, but usually containing some air pockets, formed on exposed objects by the freezing of a film of supercooled water deposited most often by **freezing rain** or **freezing drizzle**. Glaze is denser, harder, and more transparent than either **rime** or **hoarfrost**. Accumulations of glaze result in an **ice storm**.

glitter—Spots of light reflected from a point source by the surface of the sea. See **sunglint**.

global circulation—See **general circulation**.

global climate change—A **planetary-scale** variation in one or more **climatic elements** such as temperature or precipitation of a magnitude that may be sufficiently large or long-term as to seriously impact society.

global radiation—The total of **direct solar radiation** and **diffuse sky radiation** received by a unit horizontal surface. Global radiation is measured by **pyranometers**.

globe lightning—Same as **ball lightning**.

glory—An optical **diffraction** effect consisting of concentric rings of color centered about the shadow of an observer's head located on a dense collection of liquid droplets. To view a glory, the observer must be located above a warm cloud or fog bank, as on a mountain overlooking a valley or in an aircraft flying above a cloud deck, with the sun at his/her back. Also called *anticorona*.

GOES—See **Geostationary Operational Environmental Satellite**.

gradient—A **vector** quantity representing the rate of change of any **field** quantity with respect to distance, as the quantity progresses toward a maximum value, typically a three-dimensional vector quantity. In meteorology, the two-dimensional gradient is used often, and in the context of the magnitude, which is inversely proportional to the **isopleth** spacing on a **synoptic analysis chart**; the direction of the gradient is perpendicular to the isopleths and is often taken to be "downslope," or from high to low values of the field variable.

gradient wind—A hypothetical large-scale, horizontal, frictionless wind that describes a curved path; it blows parallel to curved **isobars** or **contours**. It is the result of an interaction among the horizontal components of the **pressure gradient force**, the **Coriolis effect**, and a **centripetal force**.

granules—Small bright features of the **photosphere** of the sun, covering 50%–60% of the surface. They have been likened in appearance to rice grains. Granules are about 1000 kilometers in diameter, have half-lives of about 2 minutes, and seem to be about 100°C (212°F) hotter than their surroundings. They may be caused by **convection** in the solar atmosphere.

graupel (also known as *soft hail*)—**Snow pellets**, generally 2–5 millimeters in diameter, formed in a **convective cloud** when supercooled water droplets collide and freeze on impact; a form of **frozen precipitation**.

gravitation—The **acceleration** produced by the mutual attraction of two masses, directed along the line joining their centers of mass, and of magnitude inversely proportional to the square of the distance between the two centers of mass. However, the rotation of the earth and atmosphere modifies this field to produce the field of **gravity**.

gravitational settling—Movement of particles in a **fluid** downward toward the lowest possible level under the influence of **gravity**.

gravitational water—Water in the unsaturated zone of the soil that moves under the influence of **gravity**.

gravity—The force imparted by the earth to a mass that is at rest relative to the earth. Often, the acceleration of gravity is used interchangeably with the force of gravity, since a unit mass is assumed. Since the earth is rotating, the force observed as *gravity* is the resultant of the force of **gravitation** and the **centripetal force** arising from this rotation. The net force is a **vector** directed perpendicular to sea level and to its surfaces of equal **geopotential**. The magnitude of the force of gravity at sea level decreases from the poles, where the centripetal force is zero, to the equator, where the centripetal force is a maximum and directed along the force of gravitation. This difference is accentuated by the shape of the earth, which is nearly that of an oblate spheroid of revolution slightly depressed at the poles. Also, because of the asymmetric distribution of the mass of the earth, the force of gravity is not directed precisely toward the earth's center. In many applications, a **standard gravity** is used.

gravity correction—An adjustment of the reading of a **mercury barometer** to adjust local gravity to **standard gravity**. Local gravity varies with latitude and altitude. See **barometric correction**.

gravity wave—A wave disturbance in which **buoyancy** acts as the restoring force on parcels displaced from **hydrostatic equilibrium**; hence, **gravity** is a primary control in wave propagation. A direct oscillatory conversion exists between **potential** and **kinetic energy** in the wave motion. In oceanography, a gravity wave is a water wave on the ocean surface that is larger than a **capillary wave**, since its speed of propagation is controlled primarily by gravity; typically, most water waves with a wavelength larger than 2 centimeters are gravity waves.

gravity wind(or a *drainage wind*)—A wind (or component thereof) directed down the slope of an incline and caused by greater air density near the slope than at the same levels some distance horizontally from the slope. This term usually is applied when the density difference is produced by local surface cooling along the incline, as in the case of a **mountain wind**. Usually smaller than a **fall wind**, which has an accumulation of cold air. Also known as a **katabatic wind**.

graybody—A hypothetical "body" that absorbs some constant fraction, between zero and one, of all **electromagnetic radiation** incident upon it. This fraction is called the **absorptivity**, which here is independent of wavelength. As such, a graybody represents a surface of absorptive characteristics intermediate between those of the theoretically limiting **whitebody** and a **blackbody**. No such substances are known in nature.

Great Basin high—An **anticyclone** centered over the Great Basin of the western United States. It is a frequent feature of the surface chart in the winter season.

great-circle—A locus of points formed by the intersection of the surface of a sphere and a plane passing through the center of the sphere.

great-circle course—A **course**, route, or track along a **great circle** over the earth's surface. The great-circle course is the least distance between any two points on a sphere. The angle between the great circle and **true north** changes along the course except along a **meridian** or the **equator**.

green flash—A brilliant green coloration of the upper limb of the sun occasionally observed just as the sun's apparent disk is about to sink below a distant clear horizon; considered as a **twilight phenomenon**.

greenhouse effect (also known as *atmospheric effect*)—The heating effect exerted by the atmosphere upon the earth by virtue of the fact that the atmosphere (mainly, its water vapor) absorbs and reemits **infrared radiation**. In detail, the shorter wavelengths of downwelling solar **insolation** are transmitted rather freely through the atmosphere to be absorbed at the earth's surface. This energy is reemitted upward from the surface as longwave (infrared) **terrestrial radiation**, a portion of which is absorbed by certain atmospheric constituents (e.g., **greenhouse gases**) and again emitted in both vertical directions. Some of this downwelling emitted **atmospheric radiation** reaches the earth's surface, contributing to heating of the surface. See **enhanced greenhouse effect**.

greenhouse gases—Those gas species that absorb appreciable **terrestrial radiation** and contribute to the **greenhouse effect** in the earth–atmosphere system; the main greenhouse gas is water vapor, others are carbon dioxide, ozone, methane, and nitrous oxide.

Greenland anticyclone—The atmospheric high pressure system that appears to overlie the Greenland ice cap; analogous to the **Antarctic anticyclone**.

grid—The finite collection of points, usually uniformly spaced, to which the meteorological variables used in a numerical model apply.

ground clutter—The pattern of **radar echoes** from fixed ground **targets**, such as buildings, topographic features, or vegetation near the radar unit. This type of clutter tends to hide or confuse the **precipitation echoes** returned from nearby moving or precipitation targets.

ground fog (also called *shallow fog*)—**Fog** that is an obscuring phenomenon, important to aviation. Usually this type of fog is a **radiation fog**. According to United States observation procedure, a fog with a depth of less than 20 feet (6 meters) and a *shallow ground fog* to a depth of less than 6 feet (2 meters).

ground ice—A mass of clear ice found in frozen ground, usually in the form of **permafrost** or **fossil ice**; may also be **sea ice** covered with soil. Sometimes used for **anchor ice**.

ground speed—The speed of an airborne object relative to the earth's surface. It is the magnitude of the **vector** sum of the object's **velocity** (speed and direction) with respect to the air and the **wind velocity**; or, expressed in a different manner, the algebraic sum of the aircraft's **airspeed** and the wind factor, or the magnitude of the **wind vector** component parallel to the heading of the aircraft, being positive for a **tailwind** and negative for a **headwind**.

ground streamer—An upward advancing column of high ion-density (a **streamer**) that typically ascends from a point on the earth's surface toward a descending **stepped leader** at the start of a **lightning discharge**. The ground streamer usually joins the stepped leader at about 20 meters (50 feet) above the ground, after which the heavy surge of the **return streamer** begins. Ground streamers occur because of very high electric field intensities that build directly below the descending stepped leader. Charges in the earth move in from all sides as a consequence of induction effects exerted by the stepped leader, and finally, in a sudden burst, the ground streamer shoots upward, often out of a brilliant **corona discharge** close to the surface.

ground swell—A long ocean **swell** as it passes through shallow water; it is characterized by a marked increase in height in water shallower than one-tenth of the wavelength.

ground track—A path on the earth's surface defined by an imaginary line connecting an orbiting satellite with the center of the earth. The **subsatellite point** occurs where this line intersects the earth's surface. The ground track of a **geosynchronous satellite** essentially coincides with the subsatellite point. For all other satellites, the ground track is created by the satellite's apparent path over the ground (the series of subsatellite points connected), as the earth turns on its axis and the satellite orbits overhead.

ground visibility—Same as **surface visibility**, or in aviation terminology, the **horizontal visibility** observed at the ground.

groundwater—Subsurface water that occupies the **zone of saturation**; thus, only the water below the **water table**, as distinguished from **interflow** and **soil moisture**.

groundwater depletion curve—A curve showing the release rate of water from storage in an **aquifer**.

ground-to-cloud discharge—A **lightning discharge** in which the original **streamer** processes start upward from some usually tall object on the ground; the opposite of the more common **cloud-to-ground discharge**.

group velocity—The speed of movement of the **crests** and **troughs** of a group of waves of varying **wavelengths**. See **phase velocity**. In deep water, the group velocity of **deep-water waves** is one-half the individual wave velocity, but for **shallow-water waves**, the group velocity is the same as the wave velocity.

growing degree-day unit—Difference between the daily mean temperature and a reference temperature, defined for a specific crop; usually only the positive departures are considered and accumulated over the particular crop's **growing season**.

growing season—Generally, the portion of the year during which the temperature of cultivated vegetation (i.e., the temperature of the vegetal **microclimate**) remains sufficiently high to allow plant growth. Sometimes described as the interval between the last *killing frost* of spring and the first killing frost of autumn, or as the *frost-free season*, described by the last occurrence of 32°F in spring and the first 32°F occurrence in autumn.

growler—A small floating piece of **sea ice** broken off from an **iceberg** or **floeberg**. Classified as ice with a height less than 5 meters and length less than 15 meters.

Guiana Current—An ocean current in the western equatorial Atlantic flowing northwestward along the northern coast of South America (the Guianas). The Guiana current is an extension of the **South Equatorial Current** (flowing west across the ocean between the equator and 20°S), which crosses the equator and approaches the coast of South America. Eventually, it is joined by part of the **North Equatorial Current** and becomes, successively, the **Caribbean Current** and the **Florida Current**.

Guinea Current—A warm ocean current in the central and eastern equatorial Atlantic flowing eastward along the south coast of northwest Africa into the Gulf of Guinea. The Guinea current originates as an eastward-flowing **Equatorial Countercurrent** farther west in the Atlantic.

Gulf Stream—A warm, well-defined, swift, relatively narrow, ocean current in the western North Atlantic that originates where the **Florida Current** and the **Antilles Current** begin to curve eastward from the continental slope off Cape Hatteras, North Carolina. East of the Grand Banks, the Gulf Stream meets the cold **Labrador Current**, and the two flow eastward separated by the **cold wall**. At about 40°N, 50°W, the Gulf Stream becomes **the North Atlantic Current**, which continues east-northeastward across the ocean. Sometimes the entire **Gulf Stream system** is referred to as the *Gulf Stream.*

Gulf Stream System—Collectively, the system of **western boundary currents** in the North Atlantic Ocean including the **Florida Current**, **Gulf Stream**, and **North Atlantic Current**.

gust—A sudden brief increase in the **wind speed**, typically of less than 20-second duration. It is of a more transient character than a **squall** and is followed by a lull or slackening in the wind speed. According to United States observation procedure, a gust is reported if the **peak wind speed** observed during the 10-minute observation interval has a variation of 10 knots or more between peak to lull.

gust front—The leading edge of a mass of relatively cool, gusty air that flows out of the base of a **thunderstorm** cloud (**downdraft**) and spreads along the ground well in advance of the parent thunderstorm cell; a mesoscale **cold front**, with cold temperatures in the advancing flow produced by evaporative cooling. A **shelf** or **roll cloud** may accompany the gust front, as well as **gustnadoes**.

gustnado (derived from *gust front tornado*)—A relatively weak **tornado** associated with the **thunderstorm outflow** at the leading edge of a thunderstorm cell; often found along a **gust front**. A **debris cloud** or **dust whirl** may indicate the presence of a gustnado.

gyre—A large-scale circular ocean current system that is essentially a closed circulation regime in an ocean; this system is larger than an **eddy** or **whirlpool**. Circulation may be in either direction, producing **cyclonic** or **anticyclonic** gyres.

H

haboob—A strong wind and **sandstorm** or **duststorm** in the northern and central Sudan, especially around Khartoum, where the average number is about 24 occurrences per year; these events are of short duration (about 3 hours) and occur most often during the high sun (summer) season. The name comes from the Arabic word *habb*, meaning "wind" caused by the downdraft of a **thunderstorm**.

Hadley cell—A direct thermally driven and zonally symmetric circulation first proposed by George Hadley (1685–1768), a British meteorologist, in 1735 as an explanation for the **trade winds**. It is an essentially closed circulation system, consisting of the equatorward movement of the trade winds between a latitude of about 30° and the equator in each hemisphere, with rising wind components near the equator, poleward flow aloft, and finally descending components at about 30° latitude again.

hail—A type of **frozen precipitation** in the form of balls or irregular lumps of ice, usually consisting of concentric layers of ice. **Hailstones** come in a variety of sizes and shapes. Hail is always produced by **convective clouds**, nearly always **cumulonimbus**. **Thunderstorms** that are characterized by strong **updrafts**, an abundant supply of supercooled water droplets, and great vertical development are favorable to hail formation. *Large hail*, with diameter of 3/4 inch or greater, is a criterion for a **severe thunderstorm**.

hail prevention—Elimination of large damaging **hailstones**, usually by the introduction of large quantities of a seeding agent (usually silver iodide) into a potential hailstorm.

hailstone—A single unit of **hail**, ranging in size from that of a pea to that of a grapefruit [i.e., from less than 0.5 centimeter (0.25 inch) to more than 13 centimeters (5 inches) in diameter]. Hailstones may be spheroidal, conical, or generally irregular in shape, with alternate layers composed of **glaze** and **rime**. Hailstones have a mean density of about 0.8 gram per cubic centimeter.

hailstreak—An elongated area hit by a single volume of **hail** produced by a convective storm. The elongated dimensions and orientation of the hailstreak are due to movement of this **thunderstorm** cell.

hair hygrometer—A type of an instrument (**hygrometer**) that measures **relative humidity** by means of the variation in length of a sheaf of human hairs as they adsorb atmospheric water vapor.

half-life—The time required for a system decaying at an exponential rate (such as an element in radioactive disintegration) to be reduced to one-half its initial mass or intensity.

half-tide level—The level midway between **high water** and **low water** for any series of tidal observations. This level is generally different from **mean sea level** because of the existence of semiannual and longer-period **partial tides**.

halo—A **photometeor** representing any one of a large class of atmospheric optical phenomena that appear as colored or whitish rings, arcs, pillars, or bright spots about the sun or moon when seen through an **ice crystal cloud** or in a sky filled with falling **ice crystals**. The halos exhibiting prismatic coloration are produced by **refraction** of light by the crystals and those exhibiting only whitish luminosity are produced by reflection from the crystal faces. See **haloes** of **22°** and **46°**, **parhelion**, and **sun pillar**.

halo of 22°—The **halo** phenomenon in the form of a prismatically colored circle of 22° angular radius around the sun or moon, exhibiting coloration from red on the inside to blue on the outside. Produced by **refraction** of **polychromatic** light entering a face of the ice crystal-prism and leaving the second prism face, oriented to produce an effective prism of 60° angle.

halo of 46°—A **halo** phenomenon in the form of a prismatically colored circle or incomplete arc thereof, centered on the sun or moon and having an angular radius of about 46°. The coloration is red on the inner edge to blue on the outer edge. Produced by **refraction** of **polychromatic** light as it passes through the faces of an ice crystal with an effective prism angle of 90°.

halocline—A zone in a water column experiencing larger vertical variations in **salinity**.

hard freeze—A **freeze** in which unprotected seasonal vegetation is destroyed, the ground surface is frozen solid under foot, and heavy ice is formed on small water surfaces such as puddles and water containers. It is to be distinguished from a ''hard frost'' (**black frost**).

hard rime—Opaque, granular masses of **rime** deposited chiefly on vertical surfaces by a dense supercooled **fog**. Hard rime is more compact and amorphous than **soft rime**, and may build out into the wind as glazed cones or feathers.

harmattan—A term of Arabic origin describing a dry, dust-laden northeast **desert wind** over West Africa blowing from the Sahara Desert, especially during the low sun season; associated with the northeast **trade winds**.

harmonic wave—An oscillatory wave having a **frequency** that is an integral multiple of the fundamental (lowest) frequency of the system.

haze—Suspension in the atmosphere of extremely small, dry **aerosols**, which individually are invisible to the naked eye, but collectively give the sky an opalescent appearance; sometimes used in reference to the associated reduction in **visibility**, and, hence, considered as a **lithometeor**. Haze usually indicates subsaturated air, whereas **fog** or **mist** indicates saturated conditions.

haze horizon—The top of a **haze layer** that is confined by a low-level **temperature inversion** and has the appearance of the **horizon** when viewed from above against the sky. In such instances, the true horizon is usually obscured by the haze layer.

haze layer—A layer of **haze** in the atmosphere, usually bounded at the top by a **tem-**

perature inversion and frequently extending downward to the ground. See **haze horizon**.

heading—The direction toward which an aircraft or ocean vessel is oriented. A heading may be with reference to **true north** or **magnetic north**. The heading and **course** may be different, especially in air navigation, because of **drift** (1).

headwind—A wind that opposes the intended progress of an exposed, moving object, for example, rendering an airborne object's **airspeed** greater than its **ground speed**. Opposite of a **tailwind**.

heat (sometimes called *thermal energy*)—(1) A form of **energy** transferred between systems by virtue of a difference in temperature, and existing only in the process of **energy transformation**. By the **first law of thermodynamics**, the heat absorbed by a system may be used by the system to do **work** or to raise its **internal energy**. (2) Popularly, a condition of high atmospheric temperature.

heat advisory—Issued by the National Weather Service to alert the public of **daytime heat indices** (**apparent temperatures**) of 105°F (41°C) or above for two or more consecutive days. A heat warning may be issued under extreme conditions.

heat balance—(1) An **equilibrium** state of a thermodynamic system where the sum of all incoming **heat fluxes** (sources) into the system equal the sum of all heat fluxes (sinks) out of the system; the temperature of a system is constant as a result of the balance. (2) See **planetary heat balance**.

heat budget—An inventory of all heat energy fluxes into and out of a specific system and the heat stored by the system. See **energy budget**.

heat capacity—The ratio of the **heat** absorbed (or released) by a system to the corresponding temperature rise (or fall); units are calories per Celsius degree. See **specific heat**.

heat conductivity (or *thermal conductivity*)—An intrinsic physical property of a substance, describing its ability to conduct **heat** energy through a medium with a **temperature gradient** as a consequence of molecular motion. The thermal conductivity of a material can be quantified as the amount of heat transferred across a unit area per unit time, divided by the **temperature gradient** perpendicular to the area. Materials with high thermal conductivity (e.g., silver and copper) transfer heat energy faster than other conductors.

heat engine—A system that receives **energy** in the form of **heat**, typically from an external source, and which, in the performance of an energy transformation, does **work**. The atmosphere itself is likened to such a system.

heat equator—The approximate latitude of highest mean annual surface temperature; at about 10°N.

heat flux—The time rate of transfer of **heat** across a unit area; usually expressed in units of calories per square meter per second or watts per square meter.

heat index—The **apparent temperature** that describes the combined effect of high air temperatures and high humidity levels, which reduce the body's ability to cool itself. See also **temperature-humidity index (THI)**.

heat island—An area of higher air temperatures in an urban setting compared to the temperatures of the suburban and rural surroundings. It appears as an "island" in the pattern of **isotherms** on a surface map.

heat lightning—Nontechnically, the luminosity observed from ordinary lightning too far away for its thunder to be heard.

heat low—Same as **thermal low**.

heat of fusion (or *latent heat of fusion*)—The **latent heat** associated with the phase change from a **liquid** to a **solid**. Equivalent to *latent heat of melting*.

heat of sublimation (or *latent heat of sublimation*)—The **latent heat** associated with the phase change from a **solid** to a **vapor**. Equivalent to *latent heat of deposition*.

heat of vaporization (or *latent heat of vaporization*)—The **latent heat** associated with the phase change from a **liquid** to a **vapor**. Equivalent to *latent heat of condensation*.

heat sink—A place or process that has the capacity to remove heat **energy** from the atmosphere.

heat source—A place or process that has the capacity to add heat **energy** from the atmosphere.

heat stress—The overall effect of excessive heat on an organism. Factors contributing to heat stress include air temperature, humidity, air movement, radiant energy, physical activity, physiological factors (age, health, body characteristics), and in the case of humans, clothing. In the extreme, **hyperthermia** could result.

heat stress index—A human comfort measure of the combined effect of temperature and humidity upon the ability of the human body to cool itself; the purpose of the index is to advise people of the potential danger of **heat stress**. The heat index currently used by the National Weather Service utilizes the **apparent temperature** to identify the four danger categories and related heat syndromes.

heat transfer—The exchange of **energy** in the form of **heat** between a system and its surroundings as a result of a temperature difference; this exchange can be by **radiation**, **conduction**, or **convection** in a **fluid** and/or between the fluid and its surroundings, with the energy flow being from high to low temperature regions. *Radiation* represents the transfer of radiant energy from one region to another by electromagnetic waves, with or without an intervening medium. *Conduction*, or diffusion of heat, implies the elastic impact of fluid molecules, without any net transfer of matter. *Convection* arises from the **mixing** of relatively large volumes of fluid because of the fluid motion and may be due either to local temperature inequalities (**free convection**) or to an applied pressure gradient (**forced convection**). These three processes occur simultaneously in the atmosphere, and assessment of the contributions of their various effects is often difficult.

heat wave—An extended time interval of abnormally and uncomfortably hot and usually humid weather. To be a "heat wave" such a period should last at least one day, but conventionally it lasts from several days to several weeks.

heating degree-day unit—A form of **degree-day unit** used as an indication of fuel consumption, especially for space heating; in United States usage, one heating degree-day unit is given for each degree that the **mean daily temperature** departs below the base of 65°F (where Celsius is used, the base is usually 18°C).

heavy rain (also known as **excessive precipitation**)—Rain with a rate of accumulation exceeding a specific value, for example, 7.6 millimeters per hour.

heavy sea—The condition or **state of the sea** when waves are said to be "running high."

heavy snow—Snowfall accumulating at a rate exceeding a specific value, for example, 4 inches (150 mm) or more in 12 hours.

heavy snow warning—Issued by the National Weather Service to warn the public of the potentially life-threatening **snowfalls** with accumulations exceeding specific values; typically 4 inches or more in 12 hours or 6 inches or more in 24 hours or less. Winds are expected to be light (less than 10 mph).

heavy surf advisory—Issued by the National Weather Service to inform the public that high **surf** may pose a threat to life or property.

hectare—A metric unit of area equal to 10 000 square meters or 2.47 acres.

hectopascal (hPa)—A convenient unit for reporting **atmospheric pressure** in the **SI** system, equal to 100 pascals. See **pascal**. One hectopascal is equivalent to one **millibar**.

height pattern—The general geometric characteristics of the distribution of the **altitude** (or more precisely, the **geopotential height**) of a **constant-pressure surface** as shown by **contours** on a **constant-pressure chart**.

Heiligenschein—A diffuse white ring surrounding the shadow cast by an observer's head upon **dew**-covered ground when the **solar altitude** is low and the distance from the observer to the shadow is large. This optical phenomenon is caused by the external **reflection** and **diffraction** of sunlight off the spherical dew drops.

hertz (Hz)—The accepted name for the derived unit of **frequency** in the **SI** system; 1 hertz is equal to 1 cycle per second. Named for Heinrich Rudolph Hertz (1857–1894), a German physicist who studied the wave nature of **electromagnetic radiation**.

heterosphere—The upper portion of a two-part division of the atmosphere according to atmospheric composition; the layer above the **homosphere**. The heterosphere is characterized by variation in composition, and a gradation of the mean molecular weight of constituent gases toward lower molecular weight constituents with altitude because of **diffusion**. This region starts at 80–100 kilometers above the earth's surface, and therefore closely coincides with the **ionosphere** and the **thermosphere**.

hexagonal column—One of the many forms in which **ice crystals** are found in the atmosphere, characterized by a six-sided cross-sectional area perpendicular to a long axis. These columnar crystals are typically formed by **deposition** at ambient temperatures between –5°C and –8°C. Contrast with **ice needles**.

hexagonal platelet—A small **ice crystal** exhibiting a six-sided tabular form, with a horizontal dimension at least 10 times its thickness. These platelets are typically formed by **deposition** at ambient temperatures between –10°C and –25°C.

hibernal—Pertaining to **winter**. The corresponding adjectives for **spring**, **summer**, and **fall** are **vernal**, **aestival**, and **autumnal**.

high—In meteorology, an "area of high pressure," referring to a maximum of **atmospheric pressure** (closed **isobars**) on a synoptic **surface pressure chart**, or a maximum of **geopotential height** (closed **contours**) on a **constant-pressure chart**. The term is used interchangeably with **anticyclone**.

high aloft—Same as upper-level **anticyclone**.

high clouds—The class (or étage) of clouds based on altitude, containing **cirriform** clouds: **cirrus**, **cirrocumulus**, and **cirrostratus**. The bases of these **ice crystal clouds** are found between 3 and 8 kilometers (10 000–25 000 feet) in polar regions to 6–18 kilometers (20 000–60 000 feet) in tropical regions. Contrast **middle**, **low**, and **vertically developed clouds**.

high fog—The frequent **fog** on the slopes of the coastal mountains of California, especially applied when the fog overtops the range and extends as **stratus** over the **leeward** valleys.

high-level thunderstorm—Generally, a **thunderstorm** based at a comparatively high altitude in the atmosphere, roughly 2500 meters (8000 feet) or higher. These storms form most strikingly over arid regions, and frequently their precipitation is evaporated before reaching the earth's surface. They often pose a serious wildfire threat in the Rocky Mountains.

high tide—Same as **high water**.

high water—(1) The highest water level reached during a **flood** or reservoir operation. (2) The highest water level reached by a **rising tide** during a **tide cycle**. The accepted popular term is *high tide*.

high wind warning—Issued by the National Weather Service to warn the public of the potential that sustained winds of 40 miles per hour or greater are expected to last for 1 hour or longer. This warning is also used if winds of 58 miles per hour or greater are anticipated for any time duration.

highland climate—Same as **mountain climate**.

highland ice—A thin, unbroken sheet of ice found on a relatively flat plateau; this **ice sheet** does not mask the underlying topography. Contrast with the thicker **continental glacier**.

hoarfrost (also known as ***white frost***)—A deposit of interlocking ice crystals (hoar crystals) formed by direct deposition on objects, usually those of small diameter freely exposed to the air, such as tree branches, plant stems and leaf edges, wires, poles, etc. Heavy coatings of hoarfrost are called **white frost**. Also, frost may form on the skin of an aircraft when a cold aircraft flies into air that is warm and humid or when it passes through air that is supersaturated with water vapor. Contrast with **aircraft icing**.

hodograph—The curves joining the end points of the **vectors** that represent in polar coordinates the horizontal winds at successive levels in the atmosphere (i.e., the graphical depiction of the vertical distribution of the horizontal wind).

Holocene—The recent **geologic epoch** of the **Quaternary** period extending from the end of the **Pleistocene** (approximately 10 800 years ago) to the present. This interval since the end of the **Pleistocene glaciation** represents the present **interglacial**.

homogeneous atmosphere—A hypothetical atmosphere where mass **density** is constant with altitude. Also, rarely an **adiabatic atmosphere**.

homogeneous fluid—A **fluid** with uniform density.

homosphere—The lower portion of a two-part division of the atmosphere according to the general homogeneity of atmospheric composition; opposed to the overlying **heterosphere**. The homosphere is the region with no gross change in atmospheric composition and, hence, mean molecular weight of air from the earth's surface to about 80 or 100 kilometer altitude because of **mixing**. The homosphere includes the **troposphere**, **stratosphere**, and **mesosphere**, and also the **ozonosphere** and at least part of the **chemosphere**.

hook echo—A **radar reflectivity echo** (return signal) from the precipitation shafts falling from the lower portion of a **cumulonimbus** having a curved or hooked shape. This type of echo on a conventional weather radar unit is often associated with a **mesocyclone** and often is used as the signature indicator for a **tornado**. See **tornado echo**.

horizon—One of several lines or planes used as reference for observation and measurement relative to a given location on the earth's surface, and referred generally to a horizontal direction (i.e., at right angles to the local **zenith**). Considerable contradiction exists between the nomenclatures for the several concepts of horizon. Aside from the distinctly different geological horizons (strata of earth material), two types of horizons can be identified: earth-sky horizons (*a*, *b*, and *c* below) and celestial horizons (*d* and *e* below). Meteorology is primarily concerned with the former, astronomy with the latter. Specifically, the following constitute the major variant usages, with suggested nomenclature along with other names that have been applied. (*a*) *Local horizon*: The actual lower boundary of the observed **sky** or the upper outline or terrestrial objects including nearby natural obstructions. (*b*) *Geographic horizon* (also called *apparent horizon, local horizon, visible horizon*): The distant line along which earth and sky appear to meet. In both popular usage and weather observing, this is the usual conception of horizon. Nearby prominences are said to obscure the horizon and are not considered to be a part of it. (*c*) *Sea level horizon* (also called *ideal horizon, sensible horizon, sea horizon, visible horizon, apparent horizon*): The apparent junction of the sky and the sea level surface of the earth; the horizon as actually observed at sea. This type of horizon is used as the reference for establishing times of **sunrise** and **sunset**. (*d*) *Astronomical horizon* (also called *sensible horizon, real horizon*): The plane that passes through the observer's eye and is perpendicular to the zenith at that point; or, the intersection of that plane with the **celestial sphere** (i.e., a **great circle** on the celestial sphere equidistant from the observer's zenith and **nadir**). It is the projection of a horizontal plane in every direction from the point of orientation. (*e*) *Celestial horizon* (also called *rational horizon, geometrical horizon, true horizon*): The plane, through the center of the earth, that is perpendicular to a radius of the earth that passes through the point of observation on the earth's surface; or, the intersection of that plane with the celestial sphere.

horizontal visibility—The maximum distance that an observer can see and distinguish an object lying close to the horizontal plane coincident with the observer. If the observer were standing on the ground, this distance would be the **surface visibility**. Contrast with **vertical** and **oblique visibility**.

horse latitudes—A nautical term describing the **latitude** belts over the oceans at approximately 30°–35°N and S where winds are predominantly **calm** or very light and weather is hot and dry; these zones coincide with the location of the **subtropical high-pressure belt**.

hot-wire anemometer—An instrument (**anemometer**) used for measuring **wind speed** from the variations either in temperature or in electric resistance of a metal wire that is heated by an electric current and loses more heat as the wind speed increases.

Humboldt Current—Same as **Peru Current.** Named for Alexander von Humboldt (1769–1859), a German climatologist who made temperature observations of this current.

humid air—In atmospheric thermodynamics, air that is a mixture of **dry air** and any amount of **water vapor**, up to the saturation amount.

humid climate—A **climate** type that is characterized by sufficient moisture to support abundant plant life, especially forests. Contrast **arid climate**.

humidity—(1) Generally, some measure of the water vapor content of air. (2) Popularly, same as **relative humidity**.

humilis—An abbreviated notation for **cumulus humilis**; a **cloud species** unique to the genus **cumulus**—that is, a **fair weather cumulus**.

hurricane—A severe **tropical cyclone** with maximum 1-minute sustained surface **wind speed** greater than 64 knots (74 mph or *Beaufort force 12*) in the North Atlantic Ocean, Caribbean Sea, Gulf of Mexico, and in the Eastern North Pacific off the west coast of Mexico to the International Dateline; its counterpart in the Western Pacific is the **typhoon**. For a more complete discussion, see **tropical cyclone**.

hurricane season—That segment of the calendar year that typically experiences a relatively high incidence of **hurricanes**. According to United States practice, the official hurricane season in the Atlantic, Caribbean, and Gulf of Mexico is from 1 June to 30 November; the Eastern Pacific basin the season runs from 15 May to 30 November.

hurricane-force wind—In the **Beaufort wind scale** (*Beaufort force 12*), a wind whose speed is 64 knots (74 mph) or higher. Use of this term leads to some confusion as hurricane-force winds do occur independently of **hurricanes**.

hurricane/typhoon warning—A warning that 1-minute sustained surface winds of 64 knots (74 mph) or higher associated with a hurricane or typhoon are expected in a specified coastal area within 24 hours or less. A hurricane or typhoon warning can remain in effect when dangerously high water and exceptionally high waves continue even though winds may be less than hurricane or typhoon force.

hurricane/typhoon watch—An announcement for specific areas that an incipient hurricane/typhoon condition poses a possible threat to coastal areas generally within 36 hours.

hydraulic jump—A steady-state, finite-amplitude disturbance in a channel, in which water passes turbulently from a region of (uniform) low depth (less than that of the **critical depth**) and high velocity to a region of (uniform) high depth (greater than the critical depth) and low velocity. This sudden disturbance is accompanied by energy dissipation.

hydrodynamic instability—A situation where a parcel displacement or wave motions in a **fluid** system amplify in an example of **instability**. Examples include **baroclinic** and **barotropic instabilities**.

hydrodynamics—The scientific study of **fluid** motion, to include liquids and gases. Compare with **aerodynamics**.

hydrograph—A graphical representation of some hydrological data such as **stage** or **discharge** at a point on a stream as a function of time. The most common type, the observed hydrograph, represents **river gauge** readings plotted at time of observation. Other types of hydrographs that are statistically derived from observed data include the **distribution graph** and the **unit hydrograph**.

hydrography—The study of open bodies of water (including oceans, lakes, and rivers) embracing either (*a*) their physical characteristics, from the standpoint of the hydrologist, oceanographer, or limnologist; or (*b*) the elements affecting safe navigation, from the point of view of the mariner.

hydrologic budget (also known as *hydrologic balance*)—Generally, an inventory of the relative states of inflow, outflow, and storage of water over a given area of the earth's surface, such as a **drainage basin**. Involves assessment of **evaporation**, **precipitation**, **runoff**, and **storage**.

hydrologic basin (or *watershed*)—Surface area drained by a portion or the totality of one or several given **watercourses**.

hydrologic cycle (or ***water cycle***)—Ceaseless flow of the water substance (H_2O) among oceanic, terrestrial (e.g., rivers, lakes, glaciers, groundwater), and atmospheric reservoirs.

hydrologic year—Same as **water year**.

hydrology—The scientific study of the waters of the earth, especially with relation to the effects of **precipitation** and **evaporation** upon the occurrence and character of water in streams, lakes, and on or below the land surface. In terms of the **hydrologic cycle**, the scope of hydrology may be defined as that portion of the cycle from precipitation to re-evaporation or return of the water to the seas. *Applied hydrology* utilizes scientific findings to predict rates and amounts of **runoff** (**river forecasting**), estimate required spillway and reservoir capacities, study soil–water–plant relationships in agriculture, estimate available water supply, and for other applications necessary to the management of water resources.

hydrometeor—Any product of **condensation** or **deposition** of atmospheric water vapor, whether formed in the free atmosphere or at the earth's surface; also, any water particles blown by the wind from the earth's surface. **Clouds**, **precipitation**, **dew**, and **frost** are examples.

hydrometeorology—That part of meteorology of direct concern to hydrologic problems, particularly to flood control, hydroelectric power, irrigation, and similar fields of engineering and water resources.

hydrosphere—The part of the planet earth covered by water and ice (e.g., oceans and other large bodies of water in the liquid phase); it may exclude the permanent ice, which is in the domain of the **cryosphere**. It interacts with the **atmosphere**, **cryosphere**, **lithosphere**, and **biosphere**.

hydrostatic approximation—The assumption that the atmosphere is in **hydrostatic equilibrium**.

hydrostatic equation—A mathematical expression that relates the **acceleration** due to the vertical component of the vertical **pressure gradient force** to **gravity**. This equation describes the condition of **hydrostatic equilibrium** and serves as the vertical component of the **equation of motion**.

hydrostatic equilibrium—The state of a **fluid** whose surfaces of constant pressure and constant mass (or density) coincide and are horizontal throughout. Complete balance exists between the force of **gravity** and the **vertical pressure gradient force**.

hydrostatic pressure—(1) The pressure in a **fluid** in **hydrostatic equilibrium**—that is, the downward pressure at a point due solely to the weight of fluid in a column of unit cross-sectional area. This should not be confused with **static pressure**. (2) In hydrology, a term for **soil water** involving the pore pressure.

hyetal—Of or pertaining to rain.

hygristor—An electrical instrument containing a material (such as lithium chloride) whose electrical resistance varies with atmospheric **humidity**; used in some types of recording **hygrometers** and in **radiosondes**.

hygrometer—An instrument that measures the water vapor content of the atmosphere. See **hair hygrometer** and **dewpoint hygrometer**.

hygroscopic—Pertaining to a marked ability to accelerate the **condensation** of water vapor. In meteorology, this term is applied principally to those **condensation nuclei** composed of salts that yield aqueous solutions of a very low **equilibrium vapor pressure** compared with that of pure water at the same temperature.

hygrothermograph—A recording instrument combining, on one record, the variation of air temperature and relative humidity as a function of time. The most common hygrothermograph consists of a recording **hair hygrometer** (i.e., a *hygrograph*) and a recording **deflection-type thermometer** (i.e., a *thermograph*).

hyperthermia—A dangerous overheating of an animal body above its usual temperature. This state is caused by the failure of the body to lose heat as a result of high air temperature, high humidities, or exposure to intense sunlight. See **heat stress**.

hypolimnion—The bottom layer of dense water in a thermally stratified lake located below the **thermocline.** Contrast with **epilimnion**.

hypothermia—A dangerous fall in the temperature of an animal body below the usual level. This state is brought about when the homeostatic mechanisms fail to maintain adequate production of heat under conditions of extreme or prolonged cold.

hypoxia—A deficiency of oxygen in the blood, cells, or tissues of the body, sufficient to cause psychological and physiological disturbances. A problem at high altitudes without supplementary oxygen.

hypsometric—Relating to the height above a datum point, usually **mean sea level**.

hypsometric diagram (chart)—A representation of the respective **elevations** and **depths** of points on the earth's surface with reference to **mean sea level**.

Hz—Abbreviation for **hertz**.

I

ice—The solid form of water substance (H_2O) found in the atmosphere as **ice crystals**, **snow**, **hail, ice pellets**, etc., and on the earth's surface in forms such as **hoarfrost**, **rime**, **glaze**, **sea ice**, **glacier ice**, **ground ice**, **frazil**, **anchor ice**, etc.

ice accretion—The process by which a layer of ice builds up on solid objects that are exposed to **freezing precipitation**, supercooled fog, or cloud droplets. Describes accumulation of **glaze** on surface objects or **aircraft icing** in flight.

ice age—A major interval of **geologic time** during which extensive **ice sheets** (**continental glaciers**) formed over many parts of the world. The latest instance is termed the **Quaternary Ice Age** found during the **Pleistocene** epoch.

ice cap—A domelike perennial cover of ice and snow over an extensive portion of the earth's surface, especially in polar regions; an example of an **ice sheet**.

ice cover—Ice that has formed on the surface of an otherwise open body of water (for example, a lake or river) as a result of subfreezing temperatures. An annual ice cover commences with **freeze-up** and ends with **ice-out**.

ice crystal—(1) Any one of a number of macroscopic, crystalline forms in which ice appears, including **hexagonal columns**, **hexagonal platelets**, **dendritic crystals**, **ice needles**, and combinations of these forms. (2) A type of precipitation composed of slowly falling, very small, unbranched crystals of ice, which often seem to float in the air. It may fall from a cloud or from a cloudless sky. See **ice prism**.

ice crystal theory—See **Bergeron–Findeisen Theory**.

ice desert—Any polar area permanently covered by ice and snow, with no significant vegetation.

ice fog—A type of **fog**, composed of a sufficient number of tiny suspended ice particles, 10–100 micrometers in diameter to reduce the **horizontal visibility** to less than 1 kilometer; forms at very low ambient air temperatures (typically –30°C or below) and calm conditions at high latitudes. Optical effects associated with **ice crystals** (**halo** phenomena) are sometimes observed.

ice island—A large fragment of **ice shelf** or a **tabular iceberg** found in the Arctic Ocean.

ice jam—(1) An accumulation of broken river ice caught in a narrow channel, thereby restricting the water flow and frequently producing local **floods** during a spring **breakup**. (2) Fields of lake or **sea ice** thawed loose from the shores in early spring, and blown against the shore, sometimes exerting great pressures.

ice needles (also called *ice spicules*)—Long thin ice crystals with a cross section of hexagonal shape; these differ from hexagonal columns because of their needlelike shape. Ice needles typically form by deposition at ambient temperatures from –4°C to –6°C.

ice nucleus—Any particle that serves as a nucleus in the formation of **ice crystals** in the atmosphere, used without regard to the particular physical process involved in the **nucleation**. See **freezing nucleus**.

ice pellets (also called ***sleet***)—A type of **frozen precipitation** consisting of transparent or translucent pellets of ice 5 millimeters or less in diameter. They may be spherical, irregular, or (rarely) conical in shape. Ice pellets usually bounce when hitting hard ground, and make a sound upon impact. There are two types of ice pellets: *(a)* **frozen rain**, **drizzle**, or largely melted then refrozen snowflakes; *(b)* **snow pellets** encased in a thin layer of ice.

ice point—The temperature at which a mixture of air-saturated pure water and pure ice may exist in **equilibrium** at a pressure of one **standard atmosphere**; a fiducial point for the **Celsius** and **Fahrenheit temperature scales**.

ice prisms (also called ***ice crystals***)—A fall of tiny unbranched **ice crystals** in the form of needles, columns, or plates. Their small size may make them appear to be suspended in air. They may fall from clouds or cloudless skies in cold, stable weather conditions, leading to **halo** phenomena. Also called *diamond dust* when seen glistening in sunlight.

ice sheet—An extensive mass of ice covering a continental area. Examples include **continental glaciers**, **ice caps**, and **highland ice** (glaciers).

ice shelf—A thick continuous ice formation with a fairly level surface, formed along a polar coast and in shallow bays and inlets, where it is fastened to the shore and often reaches bottom. It may grow hundreds of kilometers out to sea. It is usually an extension of **land ice**, and the seaward edge floats freely in deep water. The **calving** of an ice shelf forms **tabular icebergs** and **ice islands**.

ice shove—The push of pans of lake or **sea ice** against or upon the shoreline as a consequence of either thermal expansion of ice or by force of the wind.

ice storm—A storm characterized by a fall of **freezing precipitation**. The attendant formation of **glaze** on terrestrial objects creates many hazards.

ice storm warning—Issued by the National Weather Service to warn the public of a potentially dangerous winter weather situation with damaging accumulations of ice expected during **freezing rain** episodes; walking and driving is expected to become extremely dangerous. Significant ice (**glaze**) accumulations are usually 1/4 inch or greater. Contrast with a **freezing rain** (or **freezing drizzle**) **advisory**.

iceberg—A large mass of freshwater ice that has broken away from land and floats in the sea, or becomes stranded in shallow water; to be distinguished from **floeberg**. See **tabular iceberg**.

ice crystal cloud—A cloud consisting entirely of **ice crystals**; to be distinguished from

water clouds and **mixed clouds**. Ice crystal or **cirriform** clouds have a diffuse and fibrous appearance.

ice crystal haze—A type of very light **ice fog** composed only of airborne **ice crystals** and at times observable to altitudes as great as 6000 meters (20 000 feet). It usually is associated with precipitation of ice crystals.

Icelandic low—The semipermanent atmospheric low pressure center located near Iceland (mainly between Iceland and southern Greenland) in the North Atlantic Ocean on charts of mean sea level pressure, especially in winter. It is a principal **center of action** in the atmospheric circulation of the Northern Hemisphere.

ice-out—The breakup of **ice cover** on a water surface in spring. A phenological event in the northern United States and in Canada.

icing level—The lowest level in the atmosphere at which an aircraft in flight does, or could, encounter **aircraft icing** conditions over a given locality.

ideal gas (or *perfect gas*)—A **gas** that has the following characteristics: (*a*) it obeys **Boyle's law** and **Charles's law**, thus satisfying the **equation of state** for ideal gases; (*b*) it has **internal energy** a function of temperature alone; and (*c*) it has **specific heat** with a value independent of temperature. Dry air satisfies these requirements to a high approximation. Water vapor is less than ''ideal''; but its mass rarely exceeds 3% of that of dry air, so that the gas laws for ideal gases are universally employed in **dynamic** and **synoptic meteorology**.

IFR—The commonly used abbreviation for *instrument flight rules*; in popular aviation terminology, descriptive of the conditions of reduced **visibility** to which instrument flight rules apply; that is, **ceilings** less than 1000 feet and/or visibility less than 3 miles.

illuminance—The total luminous flux incident upon a unit area of a surface; a counterpart of **irradiance**.

illumination—The photometric process in which light shines on an object or surface.

impaction—Removal of particles from air through their impact with vegetation, structures, and other objects at the earth's surface; a natural atmospheric cleansing process.

in situ—The acquisition of data about an object or phenomenon by an instrument device immersed in and touching the medium to be measured; for example, ordinary thermometers, and barometers at a surface weather station make in situ measurements. Contrast **remote sensing**.

inadvertent climate modification—Unintentional change of **climate** due to anthropogenic activities.

inadvertent weather modification—Unintentional change of **weather** due to anthropogenic activities.

inclination—(1) In astronomy, the angular distance (in degrees) between any two planes or their poles; usually refers to the angle between an orbital plane of a celestial object (e.g., sun, planets, satellites) and a reference plane. See **inclination of**

axis. (2) In terrestrial magnetism: (also called *dip*) at any given location, the angle between the magnetic lines of force and the earth's surface. Practically, the angle between the direction that a freely suspended magnet would dip and the local **horizon** within the plane of a magnetic **meridian**. At the **magnetic pole** the **dip angle** is 90°, such that at the magnetic north pole, the compass needle would point vertically.

inclination of axis—The angle between the rotation axis of the planet or satellite and its orbital plane; for the earth, this is also known as the **obliquity of the ecliptic**.

incompressible fluid—A **fluid** whose density remains constant regardless of pressure; contrast with a **compressible fluid**. The oceans can be treated as an essentially incompressible fluid.

incus—Same as **anvil cloud**.

index of refraction—A measure of the amount of **refraction**; the ratio of the phase velocity of an electromagnetic wave of given wavelength in a vacuum to that in a particular transparent substance. It is a function of the wavelength and the physical properties of the medium.

Indian summer—A time interval, in mid- or late autumn, of unseasonably warm weather, generally clear skies, sunny but hazy days, and cool nights. In New England, at least one *killing frost* and preferably a substantial period of normally cool weather must precede this warm spell in order for it to be considered a true "Indian summer." It does not occur every year; and in some years two or three Indian summers may occur.

indicated altitude—The **altitude** read directly from a **pressure altimeter** when set to the prescribed **altimeter setting**.

indirect cell—A closed thermal circulation in a vertical plane in which warm air sinks and cold air rises; in other words, the rising motion occurs at a lower **potential temperature** than the descending motion. Such a cell forms an energy sink. Contrast with **direct cell**.

inert gas—Any one of six gases, helium, neon, argon, krypton, xenon, and radon, all of whose shells of planetary electrons contain stable numbers of electrons such that the atoms are chemically inactive.

inertia—A property of a mass exhibited by a tendency to resist any change in its rest state or in its uniform straight-line motion (constant speed and direction) unless acted upon by a **force**.

inertial flow—Flow in the absence of external forces. In meteorology, a hypothetical **anticyclonic** circular motion in a pressure field with no horizontal pressure gradient, created by the balance of **centripetal force** and the **Coriolis effect**.

inflow—The flux or flow of a substance, typically water, into a reservoir.

inferior mirage—A spurious image of an object formed below the true position of that object by abnormal **refraction** conditions along the line of sight; one of the most common of all types of **mirage**, and the opposite of a **superior mirage**. Inferior mirages occur with a strong temperature **lapse rate**, usually created over very warm surfaces such as highways or a desert.

infiltration—The seepage of water through the soil surface into a porous medium, such as the soil; or the quantity of water entering the soil.

infiltration capacity—The maximum rate at which precipitation can pass through a unit area of soil surface into the soil, for a given soil in a given condition; also the maximum amount of precipitation.

infrared radiation (**IR**)—The **electromagnetic radiation** contained in the portion of the **electromagnetic spectrum** lying in the wavelength interval from about 0.8 micrometer (**near-infrared**) to an indefinite upper boundary sometimes arbitrarily set at 0.1 millimeter (**far infrared**). At the lower limit of this interval, the infrared radiation spectrum is bounded by **visible radiation**, whereas on its upper limit it is bounded by **microwave** radiation of the type important in radar technology.

infrared satellite image—Pictures or images from **radiometers** onboard a satellite that sense **thermal** (or **infrared**) **radiation** (typically, from wavelengths of approximately 8–12 micrometers) emitted from earth and cloud surfaces of the earth–atmosphere system; because this wavelength is within the infrared water vapor "**atmospheric window**," this imagery can detect terrestrial radiation emitted directly from the earth's surface; in satellite meteorology, an **enhanced image** is often produced. Contrast with **water vapor image** and **visible satellite imagery**.

initial condition—A prescription of the state of a dynamical system at some specified time, usually at the start of the period of interest.

initial detention—Same as **surface storage**.

initialization—Determination of a consistent set of opening (*initial*) values of the variables in a numerical forecasting model using atmospheric observations.

injection temperature—The water temperature measured at the water intake to a ship's engines. This temperature is often taken as the **sea surface temperature** even though this level may be as much as 2 meters (6 feet) below the actual water surface. Compare with **bucket temperature**.

insolation (*in*coming *sol*ar rad*iation*)—In general, **solar radiation** received at the earth's surface.

instability—(1) A property of the steady state of a system such that certain disturbances or perturbations introduced into the steady state will increase in magnitude, the maximum perturbation amplitude always remaining larger than the initial amplitude. Contrast with **stability**. (2) In the atmosphere, the criterion for instability is based on using the **parcel method**; testing using either a **saturated air parcel** or an unsaturated parcel to ascertain if the parcel continues displacement after the disturbance ceased. (3) In the ocean, this condition usually refers to a vertical displacement of a water parcel.

instability line—Any nonfrontal line or band of **convective activity** in the atmosphere. This is the general term and includes the developing, mature, and dissipating stages. However, when the mature stage consists of a line of active **thunderstorms**, it is properly termed a **squall line**; therefore, in practice, instability line often refers only to the less active phases.

instrument correction—The difference between the readings of a given instrument and those of a certified standard instrument. See discussion in part a of **barometric correction** for application to a **mercury barometer**.

instrument flight rules—See **IFR**.

instrument shelter—A boxlike structure designed to protect certain meteorological instruments from exposure to direct sunshine, precipitation, and condensation, while at the same time providing adequate ventilation to ensure correct readings. See **Stevenson screen**.

instrument weather—In aviation terminology, route or terminal weather conditions of sufficiently low **visibility** to require the operation of aircraft under **instrument flight rules (IFR)**.

insulation—The prevention of the transfer of **energy** (e.g., heat or electricity) between two conductors by separating the conductors with a nonconducting material; or, the nonconducting material itself.

intensity—(1) With respect to **electromagnetic radiation**, a measure of the radiant **flux** per unit solid angle emanating from some point source. (2) In **synoptic meteorology**, the general strength of flow around an individual **cyclone** or **anticyclone** (most often applied to the former). This concept is commonly used in terms of a process, ''intensification,'' or descriptively, as an ''intense low.'' (3) In **radar meteorology**, the amount of energy returned from a **target** in a reflectivity display.

interface—A surface separating two **fluids**, representing a **discontinuity** of some fluid property, such as density, velocity, etc.; for example, the air–sea interface.

interflow—The water, derived from precipitation, that infiltrates the soil surface and then moves laterally through the upper layers of soil above the **water table** until it reaches **a stream channel** or returns to the surface at some point downslope from its point of **infiltration**.

interglacial—Pertaining to an interval of **geologic time** (tens or hundreds of thousands of years) marked by mild climate between the **glacial stages** of an **ice age**; significant melting and wasting of **glaciers** usually occur. Contrast with **interstadial**.

intermittent stream—A stream that carries water a considerable portion of the time, but which ceases to flow occasionally or seasonally because bed seepage and **evapotranspiration** exceed the available water supply. Contrast with **ephemeral** or **perennial streams**.

internal energy—A measure of the molecular activity of a system or the summation of total energies of all molecules in a specific mass; **heat**. In an ideal gas, internal energy is directly proportional to the temperature.

internal wave—A wave in **fluid** motion having its maximum **amplitude** within the fluid or at an internal boundary (**interface**).

International Geophysical Year (IGY)—By international agreement, an interval during which greatly increased observation of worldwide geophysical phenomena is taken through the cooperative effort of participating nations. Usually refers to the IGY of 1957–1958.

International Ice Patrol—An organization operated by the U.S. Coast Guard that monitors and provides warnings to mariners of dangerous **icebergs** on shipping lanes.

International System of Units (SI)—The internationally accepted and coherent form of the metric system whose **fundamental units** are the meter, kilogram, second, ampere, Kelvin, and candela.

interstadial—A warmer substage of a **glacial stage**, on the order of 1000 years duration, during which the ice retreated temporarily because the climate was cold, but not glacial. Contrast with **interglacial**.

interstitial water—Water that resides in the pores (“*interstices*”) between grains of rock or sediments.

intertropical convergence zone (ITCZ)—A narrow, discontinuous belt of convective clouds and thunderstorms paralleling the equator and marking the **convergence** of the **trade winds** of the two hemispheres; this zone shifts seasonally.

intertropical front (also known as *tropical front*)—Quasi-permanent **front** separating the Northern and Southern Hemisphere **trade winds** or marking the extreme limit of a tropical **monsoon**.

inverse square relationship—A relation describing the dependence of a **field** quantity, such as the magnitude of a force or a flux **field**, upon the reciprocal of the square of the distance between a **point source** and a field point.

inversion—In meteorology, a departure from the usual decrease or increase with altitude of the value of an atmospheric property; also, the layer through which this departure occurs (the ‘‘inversion layer’’), or the lowest altitude at which the departure is found (the ‘‘base of the inversion’’). This term almost always means a **temperature inversion**; however, a **moisture inversion** is also defined.

ion—Any atom or molecule having a net electric charge due to an imbalance of electrons. In **atmospheric electricity**, any of several types of electrically charged submicroscopic particles normally found in the atmosphere, especially in the **ionosphere**.

ionization—In **atmospheric electricity**, the **endothermic** process by which neutral atmospheric molecules or other suspended particles are rendered electrically charged chiefly by collisions with high-energy particles. Contrast with **recombination**.

ionosphere—The atmospheric shell characterized by a high **ion** density, resulting from the **photoionization** of the atmospheric gases by energetic and extremely short **ultraviolet** radiation (with wavelengths less than 100 nanometers) emitted by the sun. Its base is at an altitude of about 70–80 kilometers and it extends to an indefinite height. Some investigators subdivide the ionosphere into the **D-**, **E-**, **F_1-**, and **F_2-layers** on the basis of the vertical distribution of **free electrons**. The ionosphere influences the propagation of radio waves.

IR—Abbreviation for infrared or **infrared radiation**.

iridescent clouds—Clouds that exhibit brilliant spots, bands, or borders of colors, usually red and green, observed up to about 30° from the sun. Coloration is due to **diffraction** phenomena, with small cloud particles producing the effect. Usually thin **cirrostratus**, **cirrocumulus**, or **altocumulus** are iridescent clouds.

Irminger Current—A warm ocean current in the North Atlantic that is one of the terminal branches of **the Gulf Stream system** (part of the northern branch of the **North Atlantic Current**); it flows toward the west from near the British Isles to off the south coast of Iceland. A small portion of the water from the Irminger Current bends around the west coast of Iceland but the greater quantity turns south and becomes more or less mixed with the water of the **east Greenland Current**.

irradiance—In **radiometry**, the total incident radiant **flux** received on a unit area of a given real or imaginary surface. Units are watts per square meter.

irreversible process—In thermodynamics, changes in one or more variables of a system such that the original state of the system cannot be restored. Contrast with **reversible process**.

irrigation—The artificial application of water to land to promote the growth of crops.

isallobar—A line of equal change in **atmospheric pressure** during a specified time interval; an isopleth of **pressure tendency**. A common form is drawn for the three-hourly local pressure tendencies on a **synoptic surface chart**.

isobar—A line of equal or constant **atmospheric pressure**; an **isopleth** of pressure. In meteorology, it most often refers to a line drawn through all points of equal atmospheric pressure along a given reference surface, such as a **constant-height surface** (notably mean sea level on surface charts), the vertical plane of a synoptic **cross section**, etc.

isobaric chart—Same as a **constant pressure chart**.

isobaric process—Any thermodynamic **change of state** of a system that takes place at constant pressure.

isobath—A **contour** (or surface) of equal depth in a body of water, represented on a **bathymetric chart**; also a contour or surface of equal depth of the **water table** below the ground surface.

isochrone—A line on a chart connecting equal times of occurrence of an event. In **weather analysis**, a sequence plotted on a map of the frontal positions at several different observation times would constitute a set of isochrones.

isodrosotherm—A line on a chart connecting points of equal **dewpoint**.

isohaline—A line (or surface) connecting points of equal or constant **salinity** in water bodies or **groundwater**.

isoheight—Same as **contour** depicting vertical height of some surface above a datum plane.

isohyet—A line drawn on a chart through geographical points recording equal amounts of precipitation during a given time interval or for a particular storm.

isolated system—A system in which no transfer of mass or heat can occur across its boundaries and no **work** is done on the system by the environment. Compare with **closed system**.

isopleth—In common meteorological usage, a line of equal or constant value of a given quantity, with respect to either space or time. More specifically, a line drawn through points on a graph at which a given quantity has the same numerical value (or occurs with the same frequency) as a function of the two coordinate variables; examples include **isobars**, **contours**, etc.

isotach—A line on a given surface connecting points with equal **wind speed**.

isotherm—A line (or surface) passing through points of equal or constant temperature.

isothermal—Of equal or constant temperature with respect to either space or time.

isothermal atmosphere—An atmosphere in **hydrostatic equilibrium** in which the temperature is constant with altitude and in which, therefore, the pressure decreases exponentially upward.

isothermal layer—Any layer where temperature is constant with altitude, such that the temperature **lapse rate** is zero. Specifically, the approximately isothermal region of the atmosphere immediately above the **tropopause**.

isothermal process—Any thermodynamic **change of state** of a system that takes place at constant temperature.

ITCZ—Acronym for **intertropical convergence zone**.

J

January thaw—A spell of mild weather popularly supposed to recur each year in late January in New England and other parts of the northeastern United States; an example of a **singularity** in the climatic record of a locale.

Japan Current—Same as **Kuroshio Current**.

jet—In meteorology, a common contraction for **jet stream**.

jet streak—A region of accelerated **wind speed** along the axis of a **jet stream**; same as *jet core* or *jet maximum*.

jet stream—Relatively narrow river of very strong horizontal winds (usually 50 knots or greater) embedded in the planetary winds aloft. These jets are typically located in the upper **troposphere** above regions of strong horizontal temperature contrasts (**fronts**). Compare with **low-level jet**. Significant regions of **wind shear** are often found surrounding regions of maximum jet stream winds. Several major jet streams include the **polar jet** and the **subtropical jet**.

jet-effect wind—A wind that is accelerated through the channeling of air by some **orographic** configuration such as a narrow mountain pass or canyon; a class of **local winds**. See **canyon wind**, **mountain-gap wind**.

jet stream cirrus—A **cirriform cloud** type (especially **cirrus**) associated with an upper tropospheric **jet stream**, usually located on the equatorward side of the jet axis.

jet stream core—Same as **jet streak**.

joule (J)—The **SI** derived unit of **energy** or **work** equal to 1 newton-meter or 0.2389 calories. Named for James Prescott Joule (1818–1889), a British physicist who studied heat.

juvenile water—Water substance derived from the interior of the earth that has not previously existed as surface or atmospheric water.

K

K-index—An operational atmospheric **stability index** indicating thunderstorm potential. Low K-index values (25 and below) suggest a low potential for thunderstorm activity, whereas high K-index values (36 and above) suggest a strong potential for thunderstorm activity.

katabatic wind—A wind, usually cold, blowing down a slope. Opposite of **anabatic wind**.

kelvin (**K**)—The **SI** fundamental unit of thermodynamic **temperature**; named for Lord Kelvin (William Thompson, 1824–1907). A degree symbol is not needed. One kelvin is equivalent to one **Celsius** degree.

Kelvin temperature scale (**K**)—An **absolute temperature scale** independent of the thermometric properties of the working substance; 0K is the lowest temperature on the Kelvin scale (**absolute zero**).

Kelvin–Helmholtz billows—Roll-like circulations within a thermally stable layer containing vertical **wind shear**. They take the form of waves that curl and break into regularly spaced turbulent eddies. Named for Lord Kelvin (William Thompson, 1824–1907), a British physicist, and Hermann von Helmholtz (1821–1894), a German physicist.

Kelvin–Helmholtz wave—Wave generated in a stable, stratified atmosphere in which the vertical **wind shear** exceeds a critical value.

kilogram (**kg**)—The SI **fundamental unit** of **mass**; 1 kilogram is equivalent to 2.205 pounds.

kinematics—The branch of mechanics dealing with the description of the motion of bodies or **fluids** without reference to the forces producing the motion; **dynamics** deals with the latter. In meteorology, the analysis of the motion of **isobars** and **fronts** when treated as geometric features of the pressure field is an example of kinematic analysis, in contrast with the use of the **equations of motion** in dynamic analysis.

kinetic energy—The **energy** that a body possesses as a consequence of its motion, defined as one-half the product of its mass and the square of its speed.

kinetic theory—The derivation of the bulk properties of **fluids** from the properties of their constituent molecules, their motions, and interactions.

Kirchhoff's law—The radiation law developed by Gustav Robert Kirchhoff (1824–1887), a German physicist. This law states that at a given temperature the ratio of the **emissivity** to the **absorptivity** for a given wavelength is the same for all bodies and is equal to the emissivity of an ideal **blackbody** at that temperature and wavelength. Loosely put, this important law asserts that good absorbers of a given wavelength are also good emitters of that wavelength.

knot—The unit of speed in the **nautical system**; one **nautical mile** per hour. It is equal to 1.1508 statute miles per hour or 0.5144 meters per second.

kona—A stormy, rain-bringing wind from the southwest or south-southwest in Hawaii. It blows about five times a year on the southwest mountain slopes, which are in the lee of the prevailing northeast **trade winds** (*kona* is the Polynesian word meaning "**leeward**"). It is associated with a southward or a southeastward swing of the **Aleutian low** and the passage of a secondary depression from northwest to southeast, north of the islands.

Köppen's classification of climates—A **climatic classification** scheme developed by Wladimir Köppen (1846–1940), a German climatologist. This scheme is based upon annual and monthly means of temperature and precipitation and also takes into account the vegetation limits. It is a tool for presenting the world pattern of climate and for identifying important deviations from this pattern.

Kuroshio—A warm ocean current in the western North Pacific flowing northeastward from Taiwan to Riukiu and then close to the coast of Japan as far as latitude 35°N; part of the **Kuroshio System**. It is a density-distribution type current, and one of the swiftest of all ocean currents. The Kuroshio is the northward flowing part of the **North Equatorial Current** (which divides east of Philippines). Beyond 35°N, where it leaves the coast of Japan, it branches to form two sections of the **Kuroshio Extension**. The current is similar to the **Florida Current** of the Atlantic Ocean.

Kuroshio Countercurrent—Part of the **Kuroshio System**. Between 155° and 160°E, considerable water turns south and southwest forming part of the Kuroshio Countercurrent. It runs at a distance of approximately 700 kilometers from the coast as the eastern branch of a large whirl on the right-hand side of the Kuroshio.

Kuroshio Extension—The warm, eastward-flowing ocean current that represents the direct continuation of the **Kuroshio** (in 35°N where the Kuroshio leaves the coast of Japan), and flows eastward in two branches, the main southern branch eventually becoming the **North Pacific Current**, and the minor northern branch eventually becoming the **Aleutian Current**; part of the **Kuroshio System**.

Kuroshio System—A system of **western boundary currents** in the North Pacific Ocean including the **Kuroshio**, **Kuroshio Extension**, **North Pacific Current**, and the lesser **Tsushima Current** and **Kuroshio Countercurrent**.

L

La Niña—An episode of strong **trade winds** and unusually low **sea surface temperatures** in the central and eastern tropical Pacific. The antithesis of **El Niño**. La Niña is derived from the Spanish for girl child. La Niña and El Niño form extremes of the atmospheric trade wind fluctuations termed the **Southern Oscillation**. See **ENSO**.

Labrador Current—A cold ocean current that flows southward from Baffin Bay, through the Davis Strait, then southeastward past Labrador and Newfoundland. East of the Grand Banks, the Labrador current meets the **Gulf Stream**, and the two flow east separated by a steep **temperature gradient** called the **cold wall**.

lake breeze—A **local wind**, similar in origin to the **sea breeze** but generally weaker, blowing from the surface of a large lake onto the shores during the afternoon; it is caused by the difference in surface temperature of land and water as in the **land and sea breeze system**. In addition to area, the depth of the lake is an important factor; a shallow lake warms rapidly and is less effective as the source of a lake breeze in summer than is a deep lake. Many communities on the Great Lakes experience this phenomenon.

lake effect—Generally, the effect of any lake in modifying the weather about its shore and for some distance downwind. In the United States, this term is applied specifically to the distinct climate of the region surrounding the Great Lakes.

lake-effect snowstorm—A potentially heavy snowfall, to include **snow squalls**, occurring on the lee shore of a lake or downwind from a lake, arising as a result of the modification of a cold **air mass** during its passage over the relatively warm water on an unfrozen lake.

Lambert' law—See **Bouguer's law**.

laminar flow—A flow in which the **fluid** moves smoothly in **streamlines** in parallel layers or sheets; a nonturbulent flow.

land and sea breeze—The complete cycle of diurnal **local winds** occurring on sea coasts due to lateral differences in surface temperature of land and sea. The nocturnal **land breeze** component of the system blows from land to sea, and the daytime **sea breeze** blows from sea to land.

land breeze—A coastal breeze blowing from land to sea, caused by the temperature difference when the sea surface is warmer than the adjacent land. An example of an *offshore breeze*. Therefore, it usually blows by night and alternates with a **sea breeze**, which blows in the opposite direction by day.

land ice—Any part of the **cryosphere** containing seasonal or permanent ice cover that has been formed on land, primarily as a consequence of the freezing of intercepted precipitation. Contrast with **sea ice**. **Icebergs**, **ice shelves**, and **ice caps** are examples of land ice.

landspout—A **tornado**; **dust whirl**.

langley—A unit of energy **flux**, in terms of energy per unit area employed in radiation theory; equal to one gram-calorie per square centimeter. Named for Samuel P. Langley (1834–1906), an American solar physicist.

lapse rate—The rate of decrease of an atmospheric variable with increasing altitude, the variable is usually air temperature, unless otherwise specified.

large-scale atmospheric processes—Processes active in the atmosphere on a scale of 10 000 kilometers. The large-scale atmospheric flows are essentially nearly **hydrostatic**, nearly **geostrophic**, and wavelike in appearance (with four or five waves around the hemisphere in the middle latitudes). They exist mainly in response to the latitudinal differences in radiative heating, to the particular value of the **Coriolis parameter**, and to the spatial distribution of oceans and continents. On this scale, the curvature of the earth becomes important.

large-scale convection—Organized vertical motion on a larger scale than atmospheric **free convection** associated with **cumuliform clouds**. The widespread patterns of vertical motion in **hurricanes** or in migratory **extratropical cyclones** are examples of such convection.

laser radar—See **LIDAR**.

latent heat—The **heat** released or absorbed per unit mass by a system in a reversible, **isobaric-isothermal** phase change; to be contrasted with **sensible heat**. In meteorology, the *latent heats of vaporization* (or *condensation*), *fusion* (or *melting*), and *sublimation* (or *deposition*) of water substance are of importance. At 0°C (32°F) these are, respectively, L_V = 2.5 megajoules per kilogram (597.3 calories per gram), L_f= 0.33 megajoules per kilogram (79.7 calories per gram), and L_s = 2.8 megajoules per kilogram (677.0 calories per gram), where 1 megajoule is 1 million joules.

latitude—The geocentric angular distance (in degrees) on the earth measured to the north (positive) or south (negative) of the geographical **equator** (latitude = 0 degrees) along the **meridian** (1) of that geographic location. Lines of latitude are known as "*parallels*" of latitude, and the area between any two latitude circles is often called a "*zone*."

law of storms—Historically, the general statement of (*a*) the manner in which the winds of a **cyclone** rotate about the cyclone's center, and (*b*) the way that the entire disturbance moves over the earth's surface. The formation of this ''law'' was largely due to the investigations of H. Brandes (1826), H. Dove (1828), and W. Redfield (1831). This knowledge of the general behavior of storms led to the issuance of rules for seamen instructing them in means of navigating to avoid the dangers of storms at sea.

layer depth—In oceanography, the thickness of the **mixed layer**; or the vertical distance from the surface down to the top of the main **thermocline**.

layer of no motion—A layer, assumed to be at rest, at some depth in the ocean. This implies that the **isobaric** (pressure) surfaces within the layer are level, and hence they may be used as reference surfaces for the computation of absolute **geostrophic currents**.

LCL—Acronym for **lifting condensation level**.

lead—A navigable passage through floating **pack ice**, perhaps several kilometers wide.

leader—The **streamer** that initiates the first phase of each **stroke** of a **lightning discharge**. Like all streamers, it is a channel of very high ion density that propagates through the air (generally cloud-to-ground) by the continual establishment of an electron avalanche ahead of its tip. See **stepped leader** and **dart leader**.

lee eddies—The small irregular motions or **eddies** produced immediately in the rear (downstream) of an obstacle in turbulent **fluid**.

lee trough—Same as **dynamic trough**.

lee wave—Any wave disturbance caused by, and therefore stationary with respect to, some barrier (e.g., mountains) in the **fluid** flow; generally found downwind or on the **leeward** side of a barrier.

leeward—The direction or the side of any object (e.g., island, mountain, or region) that faces away from the prevailing wind. Opposite of **windward**.

Lenard effect—Separation of electric charges in falling rain, caused by the **aerodynamic** fragmentation of water droplets, the droplets becoming positively charged and the air negatively charged; named for Phillipp Lenard (1862–1947), German physicist.

length of record—The time interval during which observations have been maintained without interruption at a meteorological station, and which serves as the frame of reference for climatic data at that station. See **period of record**.

lenticular cloud—A common term for **lenticularis**.

lenticularis—A **cloud species** of typically the **cirrocumulus** and **altocumulus** cloud **genera** having the form of more or less isolated, generally smooth double convex lenses or almonds with sharp outlines. These clouds are most often of **orographic** origin, the result of **lee waves**, and remain nearly stationary with respect to the terrain; but they also occur in regions without marked orography as a result of **internal waves**.

level—(1) A quasi-horizontal surface, usually of constant pressure or of a function of pressure, to which the variables of a numerical model apply. (2) Altitude specified by atmospheric pressure value, as in **LFC**, **LCL**, and **CCL**.

level of free convection (LFC)—The level in the atmosphere at which an **air parcel** lifted dry-adiabatically until saturated and saturation-adiabatically thereafter would first become warmer than its surroundings in a conditionally unstable atmosphere.

level of nondivergence—The region in the **troposphere** where the horizontal **divergence** approaches zero; this diffuse level at approximately 600 hectopascals (millibars) separates the regions of divergence and **convergence** in the lower troposphere from the compensating regions in the upper troposphere. See **Dines' compensation**.

LFC—Acronym for **level of free convection**.

LIDAR (*li*ght *d*etection *a*nd *r*anging)—An instrument that emits an intense pulsed light beam (*laser*) that is reflected by tiny particles suspended in the atmosphere. In this way, the instrument measures both particle concentration and particle movement (a measure of **wind speed**).

Lifted Index (LI)—A **stability index** used operationally to determine the occurrence of **severe weather**. Stable conditions are indicated by values of LI greater than 2, very unstable conditions by LI less than –2.

lifting condensation level (LCL)—The level in the atmosphere at which an unsaturated **air parcel** lifted dry adiabatically would become saturated. See **convective condensation level** (**CCL**).

light—(1) Usually, **visible radiation** (between 0.4 and 0.7 micrometers in wavelength) evaluated in proportion to its ability to stimulate the sense of sight. (2) Of lesser intensity, for example, **light breeze**.

light air—In the **Beaufort wind scale** (*Beaufort force number 1*), a wind whose speed is from 1 to 3 knots (1 to 3 mph), causing smoke to show **wind direction** and small ripples to appear on the sea surface.

light breeze—In the **Beaufort wind scale** (*Beaufort force number 2*), a wind whose speed is from 4 to 6 knots (4 to 7 mph), causing leaves on trees to rustle and producing small **wavelets**.

light wind—According to United States observational procedure, a term used to indicate a **wind speed** of 6 knots or less.

lightning—Generally, any and all of the various forms of visible **electrical discharge** produced by **thunderstorms**; an **electrometeor**.

lightning channel—The irregular path through the air along which a **lightning discharge** occurs; may be highly ionized.

lightning detection network (LDN)—A national sensor system that provides real-time information on the location and frequency of **lightning discharges**, especially **cloud-to-ground**.

lightning discharge—An electrical charge transfer and neutralization along a narrow channel of high ion density in response to the buildup of an electrical potential between cloud and ground (**cloud-to-ground discharge**), between clouds (**cloud-to-cloud discharge**), within different portions of a single cloud (**cloud discharge**), or cloud to air (**air discharge**); an **electrometeor**.

lightning flash—In **atmospheric electricity**, the total observed luminous phenomenon accompanying a **lightning discharge**; may constitute a **composite flash**.

lightning rod—A grounded metallic conductor with its upper extremity extending above the structure that is to be protected from damage by lightning.

lightning stroke—Any one of a series of repeated **electrical discharges** composing a single **lightning discharge** (or **lightning flash**); specifically, in the case of the **cloud-to-ground discharge**, a **leader** plus its subsequent **return streamer**.

likely—In a **probability of precipitation** statement within a public forecast, a 60% or 70% chance of the occurrence of measurable precipitation.

limnology—The scientific study of lakes and open reservoirs, including hydrological phenomena, emphasizing the physical, chemical, hydrological, and ecological aspects.

line source—In air pollution studies, an extended origin of pollution from a series of points spaced sufficiently close that the **plume** may be regarded as emanating from a line; an example is a highway. Contrast with **area source** and **point source**.

liquid—One of the basic three physical **phases** of matter; a **fluid** characterized by free internal movement of the constituent molecules but without the tendency to separate. Compare with **gas** and **solid**.

liquid-in-glass thermometer—A type of indicating thermal sensor (**thermometer**), based on the principle that a **liquid**, such as alcohol or mercury, will expand or contract in a receptacle bulb and thin sealed glass tube of uniform bore as a response to ambient air temperature changes. See **alcohol** and **mercury thermometers**.

lithometeor—Anything in the atmosphere consisting of a visible concentration of mostly solid, dry particles that are lifted from the ground by the wind and/or suspended in the atmosphere. These **meteors** typically restrict **horizontal visibility** to 9 kilometers (6 miles) or less and include **blowing dust**, **haze**, **smoke**, **blowing sand**, and **volcanic ash**.

lithosphere—The outer, solid portion of the planet earth; includes the crust plus the rigid upper part of the mantle. See **atmosphere**, **hydrosphere**, **cryosphere**, and **biosphere**.

Little Ice Age—An extended time interval of relatively cold conditions in many regions of the globe from about 1400 AD to 1850. **Sea ice** cover expanded, alpine glaciers advanced, and growing seasons shortened causing much hardship to the populations of northern latitudes.

littoral zone (or *intertidal zone*)—The zone along the shoreline bounded by mean **high water** and **low water** levels produced by **tides**.

local forecast—A weather forecast made specifically for a relatively limited area, such as a city, or county/parish sized zone in contrast to a **regional forecast** for a region or state.

local wind—A wind regime of limited spatial and temporal dimensions, caused by local atmospheric conditions rather than forced by large-scale factors. Typically these winds operate on a space scale of several kilometers, between the **microscale** and **mesoscale** features, and on a timescale of several hours. Examples of local wind regimes include **land and sea breeze** and the **mountain and valley winds**.

lofting—A pollution **plume** with a flat base but appreciable vertical spread indicative of an atmosphere that is statically stable up to the base of the plume and unstable above.

longitude—The geocentric angular distance on the earth measured (in degrees) east or west along the earth's **equator** from the Greenwich **Prime Meridian** to the **meridian** (1) of the particular geographic location. Lines of longitude often are simply called *meridians*.

long-range forecast—A weather prognosis (**forecast**) for a time interval generally beyond 10 days of issuance, or beyond the **medium-range forecast**; sometimes called an *outlook*.

longshore current—Flow of water parallel to the shore and in the **surf zone**; the consequence of waves breaking at an angle to the shore.

longshore drift—Sediment transported along a beach by the **longshore current**.

long wave—(1) With regard to atmospheric circulation, a wave in the major belt of westerlies characterized by large wavelength and significant amplitude. Also known as **planetary** or **Rossby waves**. The wavelength is typically longer than that of the rapidly moving individual **cyclonic** and **anticyclonic disturbances** of the lower **troposphere**. Typically, 3–5 long waves can be found around the Northern Hemisphere at any given time. Compare with **short wave**. (2) In oceanography, a **shallow-water wave**.

longwave radiation—In meteorology, same as **infrared radiation**.

looming—A **mirage** produced by greater-than-normal **refraction** in the lower atmosphere, thus permitting objects to be seen that are usually below the local horizon. This occurs when the air density decreases more rapidly with altitude than in the normal atmosphere. Looming is the opposite of **sinking**.

looping—A pollution **plume** with large, distorted vertical eddies associated with convection in an unstable atmosphere.

low—In meteorology, an area of low **atmospheric pressure**, referring to a minimum of air pressure (closed **isobars**) on a **constant-height chart** or a minimum **geopotential height** (closed **contours**) on a **constant-pressure chart**. Same as a **cyclone**.

low aloft—Same as upper-level **cyclone**.

low clouds—The class (or étage) of clouds based on altitude, comprising **stratocumulus** and **stratus** clouds; some schemes include **nimbostratus** (otherwise, in middle clouds), **cirrus**, and **cumulonimbus** (the latter two, also vertically developed clouds). This class of clouds are found (or have bases) ranging between the earth's surface and 2 kilometers (6500 feet) in essentially all geographic regions. Contrast with **middle**, **high**, and **vertically developed clouds**.

low water—(1) The lowest water level reached by the **falling tide** during a **tide cycle**. The accepted popular term is *low tide*. (2) Lowest water level reached in a river or lake. Contrast with **high water** (1).

lower atmosphere—Generally and quite loosely, that part of the atmosphere in which most **weather** phenomena occur (i.e., the **troposphere** and lower **stratosphere**); hence, used in contrast to the common meaning for the **upper atmosphere**. In other contexts, the term implies the lower troposphere, usually below 700 hectopascals (millibars).

low-level jet—A band of strong winds (jet stream) at an atmospheric level well below the high **troposphere** as contrasted with the **jet streams** of the upper troposphere. Examples include a jet along mountains (**barrier jet**) and a **nocturnal jet**.

lull—A momentary decrease in the speed of the wind. Contrast with **peak wind** or **gust.**

luminance—A measure of the photometric **brightness** of a surface, equal to the flux of light passing through a unit area of surface and spread out over a unit solid angle; a counterpart of **radiance**.

lunar atmospheric tide—An **atmospheric tide** due to the gravitational attraction of the moon. The only detectable components are the 12-lunar-hour or **semidiurnal**, as in the oceanic tides, and two others of very nearly the same period. The amplitude of this atmospheric tide is so small that it is detected only by careful statistical analysis of a long record, being about 0.06 hectopascals (millibars) in the Tropics and 0.02 hectopascals (millibars) in middle latitudes.

lunar day—(1) The time required for the earth to rotate once with respect to the moon—that is, the time between two successive upper transits of the moon; also known as the *tidal day*. The mean lunar day is 24 hours, 50 minutes, or approximately 1.035 times greater than the **mean solar day**. (2) In astronomy, the time required for the moon to revolve once, relative to a fixed star, about its own axis.

lunar month—The interval of time or period of the complete revolution of the moon about the earth, reckoned either by phase of the moon (a *synodic month* of approximately 29.531 mean solar days) or by an absolute frame of reference (a *sidereal month* of 27.32 days).

lunar tide—That portion of a **tide** due to the **tide-producing force** of the moon. See **lunar atmospheric tide**.

lunation—The time interval required for the moon to complete all its sequence of **phases** from one new moon to the next; approximately 29.531 days; one of the means for describing the **lunar month**.

lysimeter—An instrument designed to measure the water content of the soil, the **actual evapotranspiration** rates, and **latent heat** fluxes at the surface. The instrument consists of a typical volume of soil and natural vegetation cover isolated from the surroundings to permit its weight to be continuously measured; changes in weight are related to changes in the water budget.

M

Mach—Usually associated with the number representing the ratio of the speed of an object moving through a medium to the **speed of sound** in the medium; used especially in describing *supersonic* speeds exceeding the speed of sound. Named for Ernst Mach (1838–1916), an Austrian physicist.

mackerel sky—A sky with considerable **cirrocumulus** or small-element **altocumulus** clouds, resembling the scales on a mackerel.

macroburst—A **downburst** (strong downdraft) that affects a path longer than 4 kilometers (2.5 miles) and may persist for up to 30 minutes. Surface winds may reach 210 kilometers per hour (130 miles per hour).

macroclimate—The general large-scale climate of a large area or country, as distinguished from the **mesoclimate** and **microclimate**.

macrometeorology—The study of the largest or **planetary-scale** aspects of the atmosphere, such as the **general circulation**, as distinguished from **mesometeorology** and **micrometeorology**.

magnetic field—A region of space where a dipole magnet, or an electrical conductor carrying an electric current, would experience a magnetic force or torque. The field is often described by an array of magnetic lines of force emanating from the magnet's poles.

magnetic meridian—A line that passes through both the north and south **magnetic poles**. This north–south line would parallel the orientation of a magnetic compass and corresponds to a line of equal magnetic **declination**.

magnetic north—At any point on the earth's surface, the horizontal direction of the earth's magnetic lines of force toward the north **magnetic pole**—that is, a direction indicated by the needle of a magnetic compass. Contrast with **true north**.

magnetic pole—In geomagnetism, either of the two points on the earth's surface at which the earth's magnetic lines of force converge—that is, where the **magnetic field** is vertical. Compare with **geomagnetic pole**. At present, the north magnetic pole is in the Canadian archipelago (78°N, 105°W) and the south magnetic pole is off Antarctica (64°S, 138°E).

magnetic storm—A worldwide disturbance of the earth's **magnetic field**. Magnetic storms are frequently characterized by a sudden onset, in which the magnetic field undergoes marked changes in the course of an hour or less, followed by a very gradual return to normalcy, which may take several days. Magnetic storms are caused by solar disturbances, though the exact nature of the link between the solar and terrestrial disturbances is not understood.

magnetosphere—The tear-drop-shaped region of rarefied ionized gas surrounding the earth where its **magnetic field** controls the motions of charged particles. The magnetic field presents an obstacle to the **solar wind**, as a rock in a running stream of water. This obstacle slows down, heats, and compresses the solar wind, which then flows around the rest of earth's magnetic field. As a result of this interaction, the magnetosphere extends from approximately 150 kilometers altitude out to 70 000 kilometers on the side facing the sun, but to 300 000 kilometers on the side away from the sun.

main stroke—Same as **return streamer**.

mamma (also known as *mammatus*)—Hanging protuberances, like pouches, on the under surface of a cloud. This supplementary cloud feature occurs mostly with **cirrus**, **cirrocumulus**, **altocumulus**, **altostratus**, **stratocumulus**, and **cumulonimbus**; in the case of cumulonimbus, mamma generally appear on the underside of the **anvil**. While mamma may be observed with **severe thunderstorms**, they do not produce the severe weather, nor should they be assumed as a definitive indicator of severe weather.

mandatory level—One of several **constant-pressure levels** in the atmosphere for which a complete evaluation of data derived from **upper-air observations** is required. Currently the mandatory pressure values are 1000, 925, 850, 700, 500, 400, 300, 200, 250, 100, 70, 50, 30, 20, and 10 hectopascals (millibars). To have a more complete vertical picture, **significant levels** of radiosonde observations are also evaluated. The operational **constant pressure charts** and the datasets ingested into the **operational numerical weather prediction** models correspond to many of these mandatory levels.

manometer—An instrument for measuring changes in the **pressure** of gases. A **mercury barometer** is a manometer for measuring air pressure.

mares' tails—Long, well-defined wisps of **cirrus** clouds, thicker at one end than the other.

marine climate—A regional climate under the predominant influence of the sea, characterized by relatively small seasonal variations and high atmospheric moisture content. The antithesis of a **continental climate**.

marine forecast—A forecast, for a specified oceanic and/or coastal area, of **weather elements** of particular value to maritime interests and coastal residents. These elements include wind, visibility, the general state of the weather, and storm **warnings**, when applicable.

marine meteorology—That part of meteorology that deals mainly with the study of oceanic areas, including island and coastal regions. It serves the practical needs of surface and air navigation over the oceans as well as conducting studies of air–sea interaction.

marine weather observation—The weather as observed from a ship at sea, usually taken in accordance with procedures specified by the **World Meteorological Organization**. The following elements usually are included: total **cloud amount**; **wind direction** and **speed**; **visibility**; **weather**; **pressure**; **temperature**; selected

cloud-layer data, that is, amount, type, and height; **pressure tendency**; seawater temperature; **dewpoint**; **state of the sea** (waves); and **sea ice**. Also included are the date and time, and the name, position, course, and speed of the ship.

maritime air—A type of **air mass** whose characteristics are developed over an extensive water surface and which, therefore, has the basic maritime quality of high humidity content in at least its lower levels. Contrast **continental air**.

maritime tundra—Treeless **tundra** found along many subarctic coastal belts, usually with a high proportion of arctic plants and animals far south of their normal limit.

mass—A fundamental quantity of an object, representing the measure of the amount of matter in a body, thus its **inertia**. Not to be confused with weight, or the gravitational force exerted on an object as a consequence of its mass. The **SI** fundamental unit of mass is the **kilogram**.

mauka breeze—Night winds of Hawaii of a cool and refreshing nature.

Maunder minimum—A 70-year period from 1645 to 1715 when the number of reported sunspots were relatively rare.

maximum temperature—Highest air temperature attained during a specific time interval, usually 24 hours.

maximum thermometer—A **thermometer** so designed that it registers the highest temperature attained during an interval of time before it is reset.

MCC—See **mesoscale convective complex**.

mean annual range of temperature—Difference between the mean temperature of the warmest and coldest months of the year.

mean daily temperature—The arithmetic average of the air temperature observations made at 24 equidistant times in the course of a continuous interval of 24 hours (normally the local calendar day, from midnight to midnight); or a combination of temperatures observed at less numerous times, so arranged as to depart as little as possible from the mean defined above. Commonly computed by averaging the 24-hour **maximum** and **minimum temperature**.

mean free path—The average distance traveled by the molecules of an **ideal gas** between consecutive collisions with one another.

mean sea level (**MSL**)—The average sea surface level for all stages of the **tide** over a 19-year period, usually determined from hourly heights observed above a fixed reference level; the reference surface for essentially all **altitudes**; in aviation, the level above which altitude is measured by a **pressure altimeter**. The 19-year reference period is selected to allow for essentially all possible lunar configurations associated with the **metonic cycle**.

mean solar day—The interval of time between two successive **meridional transits** of the "mean sun," an imaginary point moving with such constant **angular velocity** along the **celestial equator** as to complete one annual circuit in an elapsed time exactly equal to that of the apparent (true) sun in its annual circuit. The mean solar day is 86 400 seconds, or 1.0027379 sidereal day.

mean solar time—The time reckoned upon the constant diurnal motion of a fictitious "mean sun." Contrast with the variable **apparent solar time**, which may depart by as much as 15 minutes from mean solar time over the course of the year. See the **equation of time**.

mean temperature—The average air temperature as indicated by a properly exposed thermometer during a given time interval, usually a day, a month, or year.

Mediterranean climate—Characterized by mild, wet winters and warm to hot, dry summers; typically occurs on the west side of continents between about 30° and 45° latitude; also called *dry-summer subtropical climate*.

medium-range forecast—A weather prognosis (forecast) for a **forecast period** extending from 3 days to 10 days from the day of issue, or covering the interval between **the short-range** and **long-range forecasts**.

melting—A change in phase of water substance from solid to liquid requiring the addition of **latent heat** to the water molecules. Opposite of **freezing**.

melting level—The **altitude** at which **ice crystals** and snowflakes melt as they descend through the atmosphere. In cloud physics and in **radar meteorology**, this is the accepted term for the 0°C (32°F) constant-temperature surface. See **bright band**. Essentially the same as **freezing level** in aviation.

melting point—The temperature at which a solid substance melts—that is, changes from solid to liquid form. The melting point of a substance should be considered a property of its crystalline form only. This point is not necessarily the same as the **freezing point**, as in the case of the water substance.

meltwater—The liquid water derived directly from the melting of a **snow cover** or glacial ice.

meniscus—The upper surface of a column of liquid. The curvature of this surface is dependent upon the cross-sectional area of the liquid and the relative ability of the liquid to wet the walls of the enclosure. In the case of a mercury column enclosed in a glass container, the surface is convex, since mercury does not wet glass.

mercury barometer—A glass, mercury **manometer**, used to measure **atmospheric pressure**; operates on the principle of balancing the weight of the atmosphere in a column of unit area against the weight of a measured free-standing column of liquid mercury in an evacuated glass tube; this height of this column is a measure of the air pressure and as such, is usually reported to the public in terms of inches of mercury.

mercury thermometer—A **liquid-in-glass thermometer** containing liquid mercury; often used as the **maximum thermometer**.

meridian—(1) One-half of the **great circle** on the earth's surface passing through both geographic poles and through any given point on the planet, or in essence, a meridian of **longitude**. (2) A great circle on the **celestial sphere** passing both **celestial poles** and through the local **zenith** of any location on earth.

meridional—In meteorology, longitudinal; along a **meridian**; northerly or southerly; opposed to **zonal**.

meridional cell—A very large-scale circulation in the atmosphere or ocean that takes place in a meridional plane, paralleling a **meridian** of **longitude**; an essentially closed circulation regime is assumed. Typically, these are thermally **direct circulation** cells, driven by **convection**.

meridional circulation—An atmospheric circulation in a vertical plane oriented along a **meridian** of **longitude**. The circulation regime consists, therefore, of the vertical and the meridional (north or south) components of motion only.

meridional flow—An atmospheric flow pattern in which the winds exhibit a pronounced north–south component, roughly paralleling a **meridian** of **longitude**.

meridional transit—The time of occurrence when a celestial object passes across the observer's local **meridian** (2) on the **celestial sphere**. Local noon is defined when the sun is at meridional transit.

meridional wind—The wind or wind component along the local **meridian**, as distinguished from the **zonal wind**. In a horizontal coordinate system, with the y axis directed northward, a south wind is described as having a positive meridional wind component.

mesoanalysis—Analysis of **mesoscale** weather phenomena, such as **fronts** and cloud clusters, with scales ranging from a few to some tens of kilometers.

mesoclimate—The **climate** of a natural region of small extent, for example, valley, forest, plantation, and park. Because of subtle differences in elevation and exposure, the climate may not be representative of the general climate of the region.

mesocyclone—A vertical cylinder of **cyclonically** rotating air that develops in the updraft of a **severe thunderstorm**; a stage in the development of a **tornado**. Mesocyclones have a horizontal dimension ranging from 3 to 10 kilometers.

mesometeorology—That portion of the science of meteorology concerned with the study of atmospheric phenomena on a scale larger than that of micrometeorology, but smaller than the **synoptic scale**. Compare with **macrometeorology** and **micrometeorology**.

mesonet—A relatively dense network of meteorological observing stations that is designed to resolve **mesoscale** atmospheric conditions.

mesopause—The top of the **mesosphere** and just below the **thermosphere**. This boundary corresponds to the level of minimum temperature (approximately –90°C) at an altitude of 70–80 kilometers.

mesoscale—Dimensions of an atmospheric layer that ranges from a few kilometers to some tens of kilometers horizontally and, vertically, from the ground to the top of the **friction layer**. This scale is between **microscale** and **synoptic scale**, and includes such features as **squall lines** and **mesoscale convective complexes**.

mesoscale convective complex (**MCC**)—A persistent, nearly circular, organized cluster of many interacting **thunderstorm** cells covering an area of many thousands of square kilometers; this system can persist for hours. A mesoscale convective complex is a large, circularly organized **mesoscale convective system** (**MCS**) that must meet specific criteria for size, shape, and duration. From a satellite perspective, this system appears as a large continuous shield of **anvils**, obscuring the individual **cumulonimbus** clouds. Mesoscale convective complexes may be responsible for **severe thunderstorms**, heavy rain, and **flash flooding** in the affected areas.

mesoscale convective system (MCS)—An organized cluster of **thunderstorms** that is a feature larger than individual thunderstorms, but smaller than **synoptic-scale** systems, and can last for hours. From a satellite perspective, these systems may appear to be round or linear in shape. Mesoscale convective systems include **tropical cyclones**, **squall lines**, and **mesoscale convective complexes** (MCC).

mesosphere—The atmospheric shell between about 50 kilometers and about 70 or 80 kilometers, extending from the relative temperature maximum (the **stratopause**) at the top of the **stratosphere** to the upper temperature minimum (the **mesopause**).

METAR—An international code for **aviation weather observations**; an acronym for meteorological aviation reports.

meteor—(1) Literally, anything in the air; hence meteorology and its concern with **hydrometeors**, **electrometeors**, **lithometeors**, and **photometeors.** (2) The phenomena that accompany a body from space (a **meteoroid**) in its passage through the atmosphere—that is, the flash and streak of light, the ionized trail, etc.

meteorite—That portion of a relatively large **meteoroid** that survives its passage through the atmosphere and reaches the earth's surface.

meteorogram—A chart in which one or more **weather elements** (such as temperature, dewpoint, pressure, winds) for a given weather station are plotted against time.

meteoroid—A small particle of matter in the solar system that is observable when it enters the earth's atmosphere and is accompanied by luminous phenomena.

meteorological equator—(1) The parallel of latitude 5°N, the annual mean latitude of the **equatorial trough**. (2) The axis of the **barotropic** current that characterizes the low **troposphere** in equatorial regions. This axis is marked by the **intertropical convergence zone**.

meteorological rocket—A rocket containing an instrument package (**rocketsonde**) sent aloft to make routine upper-atmospheric observations up to a 250 000 foot (76 kilometer) altitude.

meteorological tide—Annual or semiannual changes in the water level of the oceans due to shifts in prevailing winds or seasonal changes in water temperature; distinguished from **atmospheric tide**.

meteorology—The study dealing with the phenomena of the atmosphere. This includes not only the physics, chemistry, and dynamics of the atmosphere, but is extended to include many of the direct effects of the atmosphere upon the earth's surface, the oceans, and life in general. The goals often ascribed to meteorology are the complete understanding, accurate prediction, and artificial control of atmospheric phenomena.

meteor shower—A large concentration of falling **meteoroids**. Many meteor showers are recurring phenomena and their appearance can be predicted.

meter (m)—The fundamental **SI** unit of length, where 1 meter = 3.281 feet.

methane—A **greenhouse gas**, a hydrocarbon with a chemical formula CH_4 (molecular weight is 16.043). Typical concentrations are 1500 parts per billion by volume of dry air. This gas is found in natural gas and as a byproduct of carbonaceous materials, such as petroleum, coal, bogs, and marshes; hence, the common name is "swamp gas." It is also produced by ruminating animals and termites.

metonic cycle—A time interval of essentially 19 years (235 **lunations**) marking the occurrence of essentially all phase relationships between the sun, moon, and earth. After one such cycle, the sequence of lunar phases will recur on the same day of the year. This cycle is important for tidal considerations, such as the determination of **mean sea level**.

microburst—An intense **downburst** that affects a path of 4 kilometers (2.5 miles) or less and typically has a duration of less than 10 minutes; called a **macroburst** if path were greater than 4 kilometers. Microbursts may have winds reaching 280 kilometers per hour (150 knots). Depending upon the amount of precipitation in the vicinity, the microburst can be identified as a *dry microburst* or a *wet microburst*.

microclimate—The fine climatic structure of the air space that extends from the very surface of the earth to a height where the effects of the immediate character of the underling surface no longer can be distinguished from the general local climate (**mesoclimate** or **macroclimate**).

micrometeorology—That portion of meteorology dealing with observations and explanations of the smallest-scale physical and dynamic occurrences within the atmosphere and distinguished from **macrometeorology** and **micrometeorology**. Generally, studies are confined to the **surface boundary layer** and in short timescales of less than a day.

micrometer (μm)—A unit of length equal to one-millionth of a **meter** or one-thousandth of a millimeter. One micrometer is equivalent to one thousand **nanometers**. The micrometer is a convenient length unit for measuring wavelengths of **infrared radiation**, diameters of atmospheric particles, etc.

microscale—Dimensions of an atmospheric layer ranging from a few centimeters to a few kilometers in the horizontal, and from the ground to a height of about 100 meters, where the surface loses its immediate influence in the **surface boundary layer**.

microwaves—A form of **electromagnetic radiation** with wavelengths between 0.1 and 1000 millimeters, or between **infrared radiation** and **radio** waves. Microwaves are employed in **weather radars**.

middle atmosphere—The general term representing the region of the earth's atmosphere including the **stratosphere** and **mesosphere**, extending from approximately 15 to 85 kilometer altitude. Contrast with **lower** and **upper atmosphere**.

middle clouds—The class (or étage) of clouds based on altitude, containing **altocumulus** and **altostratus**; **nimbostratus** are sometimes included. These clouds typically are found between 2 and 4 kilometers (6500–13 000 feet) in polar regions and 2 and 8 kilometers (6500–25 000 feet) in tropical regions. Contrast **low**, **high**, and **vertically developed clouds**.

middle infrared—Pertaining to the region of the spectrum with **electromagnetic radiation** with wavelengths between the **near-infrared** (2.5 micrometers) and the **far infrared** (4 micrometers).

Mie scattering—Any **scattering** produced by spherical particles without regard to the comparative size of the radiation being scattered or the diameter of the particle responsible for the scattering. Named for Gustav Mie (1868–1957), a German physicist who studied this scattering phenomenon. For example, airborne particles having the same diameter as the wavelength of **visible** solar radiation, will scatter light equally at all wavelengths, giving a milky appearance to **fog** or **aerosol**-laden skies. Contrast with **Rayleigh scattering**.

Milankovitch cycles—Systematic changes in three elements of earth–sun geometry: **precession** of the solstices and equinoxes, tilt of the earth's rotational axis, and **orbital eccentricity**; affects the seasonal and latitudinal distribution of incoming solar radiation and influences climatic fluctuation of the order of tens to hundreds of thousands of years. Named for Milutin Milankovitch (1879–1958), a Serbian astronomer and mathematician who calculated these cycles and discussed the long-term climatic implications on **ice ages**.

millibar (mb)—A pressure unit of 100 newtons per square meter (1 **hectopascal**), convenient for reporting **atmospheric pressure**; 1 millibar corresponds to 0.02953 inches of mercury.

minimum temperature—Lowest air temperature attained during a specific time interval; usually 24 hours.

minimum thermometer—A thermometer that automatically registers the lowest temperature attained during an interval of time before it is reset; an **alcohol thermometer** is often used.

mirage—A **refraction** phenomenon wherein an image of some distant object is made to appear displaced from its true position because of large vertical density variations near the surface; the image may appear distorted, inverted, or wavering. See **inferior** and **superior mirages**.

mist—Suspension in the air of microscopic water droplets or wet hygroscopic particles that reduce the **horizontal visibility** at the earth's surface to less than 9 kilometers, but to not less than 1 kilometer, the upper limit for **fog**. The **relative humidity** with mist is often less than 95%. In popular usage in the United States, same as **drizzle**.

mistral—A strong, cold, dry **katabatic wind** that flows down Alpine slopes and into the Rhone River Valley of France and then to the French and Italian Riviera on the Mediterranean coast; occurs most frequently in winter. An example of a **fall wind**.

mixed cloud—A cloud composed of both supercooled water droplets and ice crystals. Contrast with **ice crystal** and **warm clouds**.

mixed layer—(1) In oceanography, the surface layer of virtually **isothermal** water, which frequently exists above the **thermocline**. Mixing of the water in this layer is accomplished by wave action or by **thermohaline convection**. (2) In meteorology, the upper portion of the **planetary boundary layer** in which air is thoroughly mixed by **convection**.

mixed tide—Two **high waters** (2 **high tides**) and two **low waters** (2 **low tides**) per tidal day with a marked **diurnal inequality**. These tides are intermediate between **semidiurnal** and **diurnal tides**.

mixing—A random exchange of **fluid** parcels on any scale from the molecular to the largest eddy. The presumption of randomness implies that any **conservative property** within the area of mixing is equalized and the **gradient** thereof is destroyed. The process of mixing is thus irreversible; **turbulence** enlarges the scale of the mixing parcels and thereby increases the rate of mixing.

mixing depth—The vertical distance between the earth's surface and altitude to which **convection** currents can uniformly disperse **pollutants**. This upper limit is usually a **temperature inversion**.

mixing length—A mean length of travel, characteristic of a particular motion, over which an **eddy** maintains its identity; analogous to the **mean free path** of a molecule.

mixing ratio—In a system of **humid air**, the (dimensionless) ratio of the mass of water vapor to the mass of dry air usually expressed in grams per kilogram. For many purposes, the mixing ratio may be approximated by the **specific humidity**.

mock sun—Same as *sun dog*. See **parhelion**.

model—A representation in any form of an object, process, phenomenon, or system, designed to understand its behavior or to make predictions; typically models involve simplifying assumptions. Examples include physical, conceptual, mathematical, and computer models. A weather map is a graphical model; **numerical weather prediction** utilizes mathematical and computer models of various atmospheric processes; the **wave cyclone** is a conceptual model.

model atmosphere—A theoretical representation of the atmosphere, particularly of vertical temperature distribution. Examples include the **barotropic model** and **standard atmosphere**.

model output statistics (MOS)—An operational forecasting method based on statistical relationships, derived over a long period of time, between locally observed **weather elements** and output parameters of a **numerical prediction model**. The method thus takes account of local climate and bias in the model.

moderate breeze—In the **Beaufort wind scale** (*Beaufort force number 4*), a wind whose speed is from 11 to 16 knots (13 to 18 mph), causing movement of small branches and frequent **whitecaps** on water surfaces.

moderate gale—Former name for **near gale** (*Beaufort force number 7*).

moist adiabat—Same as **saturation adiabat**.

moist air—In atmospheric thermodynamics, air that is a mixture of **dry air** and any amount of water vapor, up to the **saturation** amount. Generally, air with a high **relative humidity**.

moist convection—Convection involving **humid air** that results in the formation of clouds.

moist tongue—An extension or protrusion of **humid air** into a region of relatively **dry air**. Cloudiness and precipitation are closely related to moist tongues.

moist-adiabatic lapse rate—Same as **saturation-adiabatic lapse rate**.

moisture—In meteorology, a general term usually referring to the water vapor content of the atmosphere, or to the total water substance (gaseous, liquid, and solid) present in a given volume of air. In climatology, moisture refers more specifically to quantities of **precipitation** or to **precipitation effectiveness** in plant growth.

moisture inversion—An increase of the moisture content of air with altitude, as opposed to the more common decrease in moisture with height; specifically, the layer through which this increase occurs, or the altitude at which the increase begins.

mole (mol)—A unit of mass numerically equal to the molecular weight of the substance. The *gram-mole* or *gram-molecule* is the mass in grams numerically equal to the molecular weight; for example, a gram-mole of diatomic oxygen (O_2) is 32 grams. The kilomole (1000 moles) is the **SI** fundamental unit of quantity.

momentum (or *linear momentum*)—That property of a particle representing a quantity of motion given by the product of its mass times its linear velocity; units are mass times length per time. See **angular momentum**.

monochromatic—Of or pertaining to a single **wavelength**, especially for **electromagnetic radiation**. Contrast with **polychromatic**.

monsoon—Wind in the **general atmospheric circulation**, typified by a seasonally persistent **wind direction** and by a pronounced change in direction from one season to another. The term is generally confined to a situation where the primary cause is the differential heating (changing in sense from **summer** to **winter**) between a continent and the adjacent ocean, such as the Indian monsoon of south Asia or the Arizona monsoon of the Southwest United States.

monsoon climate—Type of climate found in regions subject to **monsoons** and characterized by a dry winter and a wet summer. The monsoon climate is best developed on the fringes of the Tropics (e.g., India). The Indian subcontinent has a long winter-spring **dry season** which includes a "cold season" followed by a short "hot season" just preceding the rains; a summer and early autumn **rainy season** that is usually very wet but varies greatly from year to year; and a secondary maximum of temperature right after the rainy season.

Monsoon Current—A seasonal, eastward-flowing ocean current of the northwest Indian Ocean flowing from the Arabian Sea toward the region southwest of Sri Lanka. The Monsoon current replaces the **North Equatorial Current** and the **Equatorial Countercurrent** in summer (Northern Hemisphere), when the **southwest monsoon** forms a continuation of the southeast **trade winds**.

monsoon low—A seasonal **cyclone** found over a continent in the summer and over the adjacent sea in the winter. Examples are the lows over the southwestern United States and India in summer, and those located off lower California and in the Bay of Bengal in winter.

MOS—See **model output statistics**.

mother-of-pearl clouds—Same as **nacreous clouds**.

mountain and valley winds—A **local wind** system of diurnally varying winds along the axis of a valley, blowing uphill and upvalley by day as a **valley wind**, but downhill and downvalley by night as a **mountain wind**; they prevail mostly in calm, clear weather.

mountain climate—Climate influenced by the height factor (high elevation) and characterized by lower atmospheric pressure and by intense solar radiation rich in ultraviolet rays. Mountain climates are distinguished by the departure of their characteristics from those of surrounding lowlands. Precipitation is typically heavier on the **windward** side of a mountain barrier than on the **leeward** side (**orographic precipitation**).

mountain gap wind—A **local wind** blowing through a gap between mountains; it may be an example of a **jet-effect wind**.

mountain sickness—An illness caused by relatively low oxygen levels at high elevations especially above 3000 meters (9843 feet); symptoms include headache, shortness of breath, fatigue, insomnia, and nausea.

mountain wind—The nighttime portion of the **local mountain and valley wind** system when a downslope flow develops as a consequence of **nocturnal cooling** and cool air drains downslope off the mountain. See **katabatic wind**.

Mozambique Current—That portion of the **Agulhas Current** north of 30°S along the east coast of Africa.

MSL—Abbreviation for **mean sea level**.

mud flow—A thick and viscous flow of water, earth, and debris; may be slow or fast. Usually the result of excessive precipitation in regions where steep slopes have not been stabilized.

multiple discharge—Same as **composite flash**.

multiple scattering—Any **scattering** in which **radiation** is scattered more than once by many scattering agents (e.g., gas molecules or **aerosols**) before reaching the eye, antenna, or other sensing element.

multiple tropopause—A frequent condition in which the **tropopause** appears not as a continuous single "surface" of discontinuity between the **troposphere** and **stratosphere**, but as a series of quasi-horizontal "surfaces" that are partly overlapping in steplike arrangement. The multiple tropopause is most common above regions of large horizontal temperature contrast in the troposphere. In the more extreme conditions the components of the multiple tropopause are not distinct, and the tropopause is then just a deep zone of transition between troposphere and stratosphere.

N

nacreous clouds—Rarely seen clouds that form in the **stratosphere** at altitudes of 25–30 kilometers; resemble **cirrus** or **altocumulus lenticularis** and may be composed of ice crystals or supercooled water droplets. Also known as ***mother-of-pearl clouds*** because of their soft, pearly luster.

nadir—The point on a given observer's **celestial sphere** diametrically opposite the observer's **zenith**, that is, directly below.

nanometer (nm)—A unit of length equal to one-billionth of a **meter** or one-millionth of a millimeter. One nanometer is equivalent to 10 **Ångstroms.** The nanometer is a convenient length unit for measuring wavelengths of **ultraviolet radiation**, diameters of atmospheric particles, etc.

Nansen bottle—A device used by oceanographers to obtain subsurface samples of **seawater**.

National Ambient Air Quality Standards (NAAQS)—The maximum allowable concentration of **criteria air pollutants** in ambient air as established by the U.S. Environmental Protection Agency (EPA). Primary and secondary air quality standards were set for sulfur oxides, carbon monoxide, nitrogen dioxide, ozone, lead, and suspended particles. *Primary standards* are established to protect human health, and represent the maximum concentration tolerated by humans without ill effects. *Secondary standards* are established to protect against adverse effects of atmospheric contaminants, to include public welfare.

National Meteorological Center (NMC)—The meteorological center responsible for carrying out national functions. Responsibilities usually include collecting observations and generating **numerical weather prediction** products. In the United States, this center has been reorganized to be part of the National Centers for Environmental Prediction or **NCEP**, with many of the functions being carried out by the Hydrometeorological Prediction Center.

nautical mile—The distance unit in the **nautical system**, defined as the length of one minute of arc along any **great circle** on the earth's surface. Since this actual distance varies slightly with latitude, a nautical mile by international agreement is defined as 1852 meters (6076.103 feet or 1.1508 statute miles).

nautical system—A system for expressing distance, speed, and acceleration in which (*a*) the distance of one minute of arc along a **meridian** or **great circle** is one **nautical mile**; (*b*) a nautical mile per hour is a **knot**; (*c*) a nautical mile per hour per hour is the acceleration in knots per hour.

nautical twilight—The interval of incomplete darkness before **sunrise** or after **sunset** when the time at which the center of the sun's disk is between 6° and 12° below the astronomical **horizon**, or between **civil** and **astronomical twilight**. When the sun is 12° below the horizon, sufficient light is just available to distinguish outlines of ground objects.

NCEP—Acronym for National Centers for Environmental Prediction. There are seven centers: Environmental Modeling Center, Hydrometeorological Prediction Center, Storm Prediction Center, Marine Prediction Center, Tropical Prediction Center, Aviation Weather Center, Climate Prediction Center.

neap range—The average semidiurnal **tidal range** occurring at the time of **neap tide**.

neap tide—A small-amplitude ocean tide of minimum **tidal range** occurring semimonthly near the time when the moon is in quadrature—that is, the first and third quarters.

near gale (also known as ***moderate gale***)—In the **Beaufort wind scale** (*Beaufort force number 7*), a wind whose speed is from 28 to 33 knots (32 to 38 mph), causing whole trees to move and white foam to appear on water surfaces from breaking waves.

near-infrared—Pertaining to the portion of the spectrum containing **electromagnetic radiation** lying just beyond **visible light**, having a relatively short wavelength, between 0.75 and 2.5 micrometers.

nearshore circulation—The movement of water caused by **nearshore currents**, **coastal currents**, and the **upwelling** of water.

nearshore current—A water current found shoreward of the **coastal current** system.

neoglaciation—The renewal of glacial expansion in mountain ranges after earlier shrinkage or disappearance.

nephanalysis—The analysis of a **synoptic chart** or **satellite image** in terms of the types and amount of clouds and precipitation. Cloud systems (or *nephsystems*) are identified both as entities and in relation to the **pressure pattern**, **fronts**, etc.

nephology—The study of clouds.

nephometer—A general term for instruments designed to measure the amount of **cloudiness**.

net radiation—At a given level or surface, the difference between the downward and upward fluxes of the total (**solar** plus **terrestrial**) radiation incident and reflected or scattered.

net solar radiation—Difference between the **solar radiation** fluxes directed downward and upward; net flux of solar radiation.

net terrestrial radiation—Difference between the downward and upward **terrestrial** (or long wave) **radiation** fluxes; net flux of terrestrial radiation.

neutral stability—The state of an unsaturated or saturated column of air in the atmosphere when its **environmental lapse rate** of temperature is equal to the **dry-adi-**

abatic lapse rate** or the **saturation-adiabatic lapse rate**, respectively. Under such conditions an **air parcel** displaced vertically will experience no buoyant acceleration.

Newton (**N**)—The **SI** derived unit of **force** defined as giving a 1 meter per second squared **acceleration** to a mass of 1 kilogram. Named for Sir Isaac Newton (1642–1727), the English physicist and mathematician who formulated **Newton's laws of motion**.

Newton's laws of motion—A set of three fundamental postulates forming the basis of the mechanics of rigid bodies, and first formulated by Sir Isaac Newton in 1687. The *first law* is concerned with the principle of **inertia** and states that if a body in motion were not acted upon by an external **force**, its **momentum** remains constant (law of **conservation of momentum**). The *second law* asserts that the rate of change of momentum (**acceleration**) of a body is proportional to the force acting upon the body and is in the direction of the applied force. A familiar statement of this is the equation, $F = ma$, where F is **vector** sum of the applied forces, m the mass, and a the vector acceleration of the body. The *third law* is the principle of action and reaction, stating that for every force acting upon a body, a corresponding force of the same magnitude is exerted by the body in the opposite direction.

nimbostratus (**Ns**)—A principal **cloud type** (**cloud genus**), gray colored and often dark, rendered diffuse by more or less continuously falling rain, snow, ice pellets (sleet), etc., and *not* accompanied by lightning, thunder, or hail; usually considered to be a **middle cloud**. In most cases the precipitation reaches the ground, but not necessarily, as evident by **virga**. The cloud is thick enough throughout to blot out the sun. Low ragged clouds (**scud**) frequently occur below the layer, and may or may not merge with it.

nitrogen (N_2)—A colorless and odorless gas that is the most abundant species by volume in the **homosphere**; it constitutes 78.084% by volume of **dry air**. Under normal atmospheric conditions, it is a diatomic element (molecular formula is N_2), with a molecular weight of 28.0134; monatomic nitrogen is found in relatively low concentrations, even at high altitudes, because diatomic nitrogen is difficult to dissociate. Although free atmospheric nitrogen is not exceedingly reactive, it does undergo chemical reactions as part of the **nitrogen cycle**.

nitrogen cycle—A major biogeochemical cycle in which **nitrogen** circulated through the air, soil, water, and biosphere. Free nitrogen gas in the air is converted by *bacteria (nitrogen fixation)* into substances that green plants can absorb from the soil; movement of these nitrogen compounds occurs through the food chain, with ultimate decay. Nitrogenous substances in decomposed organic matter ultimately return to the soil and atmosphere.

NMC—See **National Meteorological Center**.

noctilucent clouds—Wavy thin clouds resembling **cirrus**, but which are usually bluish white or silvery. They are best seen at high latitudes just before **sunrise** or just after **sunset**. They occur in the upper atmosphere at altitudes between 75 and 90 kilometers, are rarely seen, and may be composed of ice deposited on meteoric dust.

nocturnal cooling—Lowering of air temperature caused by nighttime or **nocturnal radiation** away from that layer of air.

nocturnal jet—Nighttime layer of strong wind at an altitude of a few hundred meters above the ground. Such a layer may develop when the strong **radiational cooling** over land at night separates the flow aloft from the constraint of **surface friction**.

nocturnal radiation—Since radiation at night has no solar flux components, this flux is the same as **effective terrestrial radiation**. (Use of this term is well established in the literature, but the current trend is toward the less misleading term, effective terrestrial radiation.)

node—A point in an oscillatory system exhibiting no oscillatory motion; especially associated with wave interference. In a **standing wave**, vertical motion is minimal at a node.

noise—Types of error in data produced by imperfections in observing techniques (observational noise) or by fluctuations on smaller time- or space scales than those being processed (meteorological noise).

nonadiabatic process—Same as **diabatic process**.

nonrecording rain gauge—An indicating instrument that measures the accumulated **catch** of precipitation, especially rain, but does not make an autographic trace of rain when it fell as does a **recording rain gauge**. This type of gauge must be read directly on a regularly scheduled daily basis. See **standard rain gauge**.

nor'easter—Common contraction for **northeaster**.

normal—The average value of a **weather element** (e.g., temperature, precipitation, humidity) over a uniform and relatively long interval (or **period of record**) covering at least three consecutive 10-year periods, such as those defining the **climatological standard normals**.

normal water—A standard **seawater** preparation, the **chlorinity** of which lies between 19.30 and 19.50 per mille (psu) and has been determined to within ±0.001 per mille. Normal water is used as a convenient comparison standard for chlorinity measurements of seawater samples by titration. It is prepared by the Hydrographical Laboratories, Copenhagen, Denmark.

North Atlantic Current—A part of the **Gulf Stream system**, representing a continuation of the **Gulf Stream**, originating at about 40°N and 50°W. It comprises all the eastward- and northward-flowing currents of the North Atlantic originating in the region east of the Grand Banks, extending toward the **Norway Current** and to the **Irminger Current**. The branches of the North Atlantic Current are often masked by shallow and variable, wind-driven surface movements so that they are sometimes called the *North Atlantic drift*.

North Cape Current—A warm extension of the **Norway Current** that flows northeastward around North Cape (the northern tip of Norway) before entering the Barents Sea in the Arctic Ocean.

North Equatorial Current—Any of several ocean currents driven by the northeast **trade winds** blowing over the tropical oceans of the Northern Hemisphere. In the

Atlantic Ocean, it flows west between the **Equatorial Countercurrent** and the **Sargasso Sea** (30°N). Part passes along the northeast side of the West Indies as the **Antilles Current** while part joins the **Guiana Current** and enters the Caribbean Sea and the Gulf of Mexico as the **Caribbean Current**. In the Pacific Ocean, it crosses from east to west between the approximate latitudes of 10° and 20°N. East of the Philippines it divides, part turning south to join the Equatorial Countercurrent, and part going north to form the **Kuroshio**. In the Indian Ocean, in northern winter when the northeast **monsoon** duplicates the **trade winds** of the other oceans, the **North Equatorial Current** flows west and turns to the southwest along the coast of Somali as the **Somali Current**. In Northern Hemisphere summer, when the **southwest monsoon** forms a continuation of the southeast trade winds, the **North Equatorial Current** and the Equatorial Countercurrent are replaced by an eastward-flowing **Monsoon Current**.

North Pacific Current—The warm southern branch of the **Kuroshio Extension** (to the east of 160°E) flowing eastward across the Pacific Ocean between 25° and 45°N; a part of the **Kuroshio System**. The main part of the North Pacific Current does not extend across the Pacific Ocean but turns back toward the west in the longitude of the Hawaiian Islands.

northeast storm—A **cyclonic** storm of the east coast of North America, so called because the winds over the coastal area preceding the storm's passage are from the northeast. They may occur at any time of year but are most frequent and most violent between September and April.

northeast trades—The **trade winds o**f the Northern Hemisphere, with a prevailing **wind direction** from the northeast. This belt is usually located between 0° and 30°N, on the equatorward flank of the **subtropical highs** of the Northern Hemisphere.

northeaster (or *nor'easter*)—A northeast wind (from a northeast direction), particularly a strong wind or **gale**; a **northeast storm** over the east coast of North America.

norther—A northerly wind coming from the north and usually with exceptional consequences. In the Southern Plains of the United States, especially in Texas, a cold, strong northerly wind that occurs especially in winter, which causes a sudden and drastic drop in air temperature. This event occurs with a cold **polar outbreak**, which spreads south into the Gulf of Mexico and occasionally reaches Central America.

northern lights—Same as **aurora borealis**.

Norway or **Norwegian Current**—Part of the northern branch of the **North Atlantic Current** that flows northward along the coast of Norway, splitting into the **North Cape Current**, which flows northeastward, and the northwestward-flowing **Spitzbergen-Atlantic Current**.

Norwegian cyclone model—The original description of the structure and life cycle of a midlatitude atmospheric low pressure system (a **cyclone**), first proposed following World War I by researchers at the Norwegian School of Meteorology at Bergen. See **polar-front theory**.

Notos—The ancient Greek name for the south wind. As a sultry and rainy wind, it is represented on the **Tower of the Winds** in Athens by a lightly clad young man carrying a jar that is inverted to allow the flow of water.

nowcasting—A description of current weather and a short-term weather forecast varying from minutes to a few hours; typically shorter than most operational **short-range forecasts**.

Ns—Abbreviation for the cloud type **nimbostratus**.

nuclear winter—A theory that a massive nuclear war would produce catastrophic **climate change** through global cooling due to blockage of incoming sunlight by extensive smoke and ash clouds.

nucleation—Any process by which the phase change of a substance to a more condensed state (**condensation**, **deposition**, **freezing**) is initiated at certain loci (see **nucleus**) within the less condensed state.

nucleus—In physical meteorology, a particle of any nature upon which, or the locus at which, molecules of water or ice accumulate as a result of a phase change to a more condensed state; an agent of **nucleation**, such as **condensation** or **freezing nuclei**.

number density—The number of entities (e.g., molecules, raindrops, etc.) per unit volume. Not to be confused with *mass density*, as described in **density**.

numerical forecasting—The forecasting of the behavior of atmospheric disturbances by the numerical solution of the governing fundamental equations of **hydrodynamics**, subject to observed **initial conditions**; electronic computers and sophisticated computational models are required.

numerical weather prediction (**NWP**)—Same as **numerical forecasting**.

O

objective analysis—A technique for the interpretation and drawing of **isopleths** on a two-dimensional display of **field** properties, utilizing sophisticated mathematical interpolation techniques; this analysis process is typically performed by electronic computers as compared with the personal **subjective analysis** techniques used by humans.

objective forecast—A weather forecast based solely upon the application of thermodynamic, dynamical, and statistical equations, without personal judgment of the forecaster.

oblique visibility—The maximum distance that an observer can see and identify an object not situated on the same horizontal or vertical plane as the observer is situated; critical for aviation interests. Compare with **horizontal** and **vertical visibilities**.

obliquity of the ecliptic—The **inclination of axis** for the earth, defined as the angle between the planes of the earth's **equator** and its orbital plane (the **ecliptic**). This orbital element determines the amplitude of the seasonal variation of incident **solar radiation**. At present the value of the obliquity is 23°26', and it undergoes a regular periodic variation from 22°3' to 24°27' with a period of approximately 41 000 years. See **Milankovitch cycles**.

obscuration—In United States weather observing practice, the designation for the **sky cover** when the sky is completely hidden by surface-based **obscuring phenomena**.

obscured sky cover—Same as **obscuration**.

obscuring phenomena—In United States weather observing practice, any **atmospheric phenomenon** exclusive of clouds that restricts **vertical visibility**. **Dust**, **rain**, **snow**, and **blowing snow** are examples.

observation—See **weather observation**.

observational error—The difference between the true value of some quantity and its observed value. Every observation is subject to certain errors, as follows: (*a*) *Systematic errors* affect the whole of a series of observations in nearly the same way. For example, the scale of an instrument may be out of adjustment. (*b*) *Random errors*, which appear in any series of observations, are generally small and as likely to be positive as negative.

obstruction to vision—In United States weather observing practice, any **atmospheric phenomenon** exclusive of weather (precipitation) phenomena that restricts **prevailing horizontal visibility** at the surface to 9 kilometers or less. **Dust**, **smoke**, **haze**, and **blowing snow** are examples.

occluded cyclone—Any **cyclone** (or low) within which an **occluded front** has developed.

occluded front (also known as ***occlusion***)—A composite of two **fronts**, formed as a **cold front** overtakes and merges with a **warm front** or **quasi-stationary front**. This is a common process in the late stages of wave-cyclone development, but is not limited to occurrence within a wave cyclone. Two basic types of occluded fronts (**cold occlusion and warm occlusion**) are determined by the relative coldness of the air behind the original cold front to the air ahead of the warm (or stationary) front.

occlusion—(1) The process of the formation of an **occluded front**; often associated with the life cycle of a **wave cyclone** according to the Norwegian model. See **cold** and **warm occlusions**, as to the types of occlusions dependent upon the relative positions of the airmass types. (2) Also used for **occluded front**.

ocean—(1) The intercommunicating body of **saltwater** occupying the depressions of the earth's surface. (2) One of the four major primary subdivisions of the above, bounded by continents, the equator, and other imaginary lines.

ocean basin—A large circular or oval-shaped depression on the ocean floor, often with **depths** exceeding 600 feet (approximately 200 meters).

ocean current—A movement of ocean water characterized by regularity, either of a cyclic nature, or more commonly as a continuous stream flowing along a definable path. Three general classes, by cause, may be distinguished: (*a*) **geostrophic currents** related to **seawater** density gradients; (*b*) **wind-driven currents**, produced by the **stress** exerted by the wind upon the ocean surface; (*c*) currents produced by long-wave motions. The latter are principally **tidal currents**, but may include currents associated with **internal waves**, **tsunamis**, and **seiches**. The major ocean currents are of continuous, streamflow character, and are of first-order importance in the maintenance of the earth's **planetary heat balance**.

oceanic climate—Same as **marine climate**.

oceanicity—The degree to which a point on the earth's surface is in all respects subject to the influence of the sea. Opposite of **continentality**.

oceanography—The study of the **sea**, embracing and integrating all knowledge pertaining to the sea's physical boundaries, the chemistry and physics of **seawater**, and marine biology.

offshore wind—A wind that blows from the land out over a body of water; the **land breeze** is an example. Opposite of an **onshore wind**.

old snow—Deposited snow in which the original crystalline forms are no longer recognizable, such as **firn** and **spring snow**.

omega high—A **ridge** in the westerly flow in the middle or upper **troposphere** that has roughly the shape of the Greek letter omega; persistent dry, fair weather under this ridge pattern is typical. Frequently becomes a **blocking high**.

onshore wind—A wind that blows from over a body of water onto the land; **sea** and **lake breezes** are examples. Opposite of an **offshore wind**.

open cells—A **mesoscale** organization of **convection** in the form of ringlike patterns of **convective clouds** as seen from a satellite perspective; these systems are usually a few tens of kilometers in diameter. Ascent of air occurs within the cell walls and descent in the cloud-free centers. Contrast with **closed cells**.

operational weather limits—The limiting values of **ceiling**, **visibility**, and wind, or **runway visual range**, established as safety minima for aircraft landing and take-off operations. Civil aircraft operate under limits stated in *Civil Air Regulations* and military aircraft operate under limits established by the respective military organizations. Limits for day and night operations usually differ. Also the limits vary according to airport environment, navigational aids, and type of aircraft.

opposing wind—Generally, same as **headwind**; specifically, a wind blowing in the direction opposite to ocean-wave advance. Opposite of a **following wind**.

orographic—Of, pertaining to, or (frequently in meteorology) caused by mountains.

orographic cloud—A cloud whose form and extent is determined by the disturbing effects of high terrain (orography) upon the passing flow of air. Because these clouds are linked with the form of the terrestrial relief, they generally move very slowly, if at all, although the winds at the same level may be very strong; often times, these are **standing clouds**. Examples include **lenticular clouds** and **banner** or **cap clouds**.

orographic lifting—The lifting of an air current by its passage up and over mountains with attendant **adiabatic cooling**; responsible for **orographic clouds** and **orographic precipitation**.

orographic occlusion—An **occluded front** in which the **occlusion** process has been hastened by the retardation of the warm front along the **windward** slopes of a mountain range, permitting the original **cold front** to overtake the **warm front**.

orographic precipitation—Precipitation that results from the lifting of **humid air** over a topographic (**orographic**) barrier such as a mountain range. Strictly, the amount so designated should not include that part of the precipitation that would be expected from the dynamics of the associated weather disturbance, were the disturbance over flat terrain.

orographic rainfall—As commonly used, same as **orographic precipitation**.

orography—The nature of a region with respect to its elevated terrain.

oscillation—Generally, the process of regularly varying above and below a mean value; usually, a periodic process.

outer atmosphere—Very generally, the tenuous atmosphere at a great distance from the earth's surface; possibly best used as an approximate synonym for **exosphere**.

outflow—(1) The **flux** or flow of a substance out of a reservoir; typically water. (2) See **thunderstorm outflow**.

outgassing—Release of gases to the atmosphere from hot, molten rock during volcanic activity; thought to be the origin of most atmospheric gases of the earth's secondary atmosphere.

outlook—A statement issued by the National Weather Service with sufficient lead time to provide useable information on the possible development of a particular weather event. Daily outlooks of severe weather potential are issued for the following 24 hours. Tropical outlooks are issued for the ocean basins discussing the conditions associated with potential tropical storm and hurricane formation. The system of **watches** and **warnings** are intended for more immediate weather situations. See **forecast**.

overcast—(1) An official **sky cover** classification for **aviation weather observations**, descriptive of a sky cover of 1.0 (95% or more) when at least a portion of this amount is attributable to clouds or **obscuring phenomena** aloft; that is, when the total sky cover is not due entirely to surface-based obscuring phenomena (see **obscuration**). (2) Popularly, the cloud layer that covers most or all of the sky. It generally suggests a widespread layer of clouds often considered typical of a **warm front**.

overland flow (also known as *surface flow*)—Water flowing over the ground surface toward a definite **channel**. Upon reaching the channel, it is called **surface runoff**.

overrunning—A condition existing when an **air mass** is in motion aloft above another air mass of greater density at the surface. This term usually is applied in the case of warm air ascending the surface of a **warm front** or **quasi-stationary front**. Generally descriptive of the precipitation so produced.

overshooting top—Protuberances from the top of a **cumulonimbus cloud** that rise above the general upper level of the cloud or **anvil** before sinking back, their appearance is that of cupolas.

overturn—The reversal of layers to include the renewal of bottom water that occurs annually in thermally stratified lakes and ponds in regions wherever winter temperatures are sufficiently cold. As the surface waters are cooled in the autumn and early winter, they become denser and therefore sink, until the whole body of water is at 4°C (39.2°F), the temperature of maximum density. Further cooling is restricted to the surface layers, since both ice and water colder than 4°C (39.2°F) are less dense than the underlying waters at 4°C (39.2°F).

oxygen (O_2)—In its free form, a colorless, tasteless, and odorless gaseous element; the second most abundant gas in the earth's atmosphere and a prerequisite of virtually all forms of animal life. It constitutes 20.9476% by volume of **dry air** in the **homosphere**. Under normal atmospheric conditions, it is a diatomic element (molecular formula is O_2) with a molecular weight equaling 31.999. Oxygen is present primarily in the molecular form up to an altitude of about 100 kilometer. Above that altitude, **photodissociation** into atomic oxygen (O) becomes more and more important so that by 150 kilometers, oxygen occurs primarily in the atomic form.

Oyashio—A cold ocean current flowing from the Bering Sea southwestward along the coast of Kamchatka, past the Kuril Islands, continuing close to the northeast coast of Japan and reaching nearly 35°N. The Oyashio turns and continues eastward, eventually jointing the **Aleutian Current**.

ozone (O_3)—An almost colorless, gaseous form of oxygen with an odor similar to weak chlorine. A relatively unstable compound of three atoms of oxygen (molecular formula O_3 and molecular weight 47.9982), ozone constitutes, on average, less than one part per million by volume (*ppmv*) of the gases in the atmosphere. Ozone is produced naturally in the middle and upper stratosphere through **photodissociation** of molecular oxygen by solar ultraviolet radiation. Peak ozone concentration occurs at altitudes between 20 and 30 kilometers (corresponding to the **stratosphere** or **ozonosphere**), with concentrations reaching as much as 10 ppmv. Its formation through photodissociation of oxygen at approximately 50 kilometers is responsible for the relative temperature maximum found at the **stratopause**. Ozone in the **troposphere** is largely a byproduct of photochemical processes associated with air pollution; it is an important constituent of **photochemical smog**.

ozone hole—A significant thinning of stratospheric **ozone** over Antarctica during the Southern Hemisphere spring; deepening of the ozone hole since the late 1970s appears linked to **chlorofluorocarbons** (**CFCs**). A less significant thinning also occurs in the Arctic.

ozone layer—Generally, any layer in the atmosphere with a maximum of **ozone** concentration; may describe the **ozonosphere**.

ozonosphere—The general stratum of the **upper atmosphere** experiencing an appreciable **ozone** concentration and in which ozone plays an important part in the radiative balance of the atmosphere. This region lies roughly between 10 and 50 kilometers, with maximum ozone concentration at about 20–25 kilometers; this layer coincides with the **stratosphere**.

P

Pacific air—A North American **air mass** originating over the Pacific Ocean that travels through the western mountains and emerges on the Great Plains warmer and drier than over its **source region**.

Pacific high—The nearly permanent atmospheric **subtropical high** pressure cell of the North Pacific Ocean, centered, in the mean, north of the Hawaiian Islands at 30°–40°N and 140°–150 °W. Considered as one of the **centers of action** in the atmospheric circulation of the Northern Hemisphere.

pack ice—Any area of **sea ice** formed by the jamming or crushing together of pieces of floating ice, resulting in a rough ice mass; the mass covers the sea surface with little or no open water.

paleoclimate—Climate of some prehistoric interval whose main characteristics may be inferred, for example, from geological and paleobiological (fossil) evidence.

paleoclimatology—The study of past climates throughout **geologic time** (**paleoclimates**), and the causes of their variations. Inferences about **climatic elements** from the fossils and character of the rocks makes paleoclimatology a branch of geology. Coordination of these inferences in terms of the effect of changes in land and sea distribution, orography, solar radiation, and astronomical factors makes it a branch of climatology.

pancake ice—Nearly circular pieces of newly formed **sea ice**, usually 2 meters in diameter, with raised rims due to pieces continually striking against each other.

parallax—The change in apparent position of a nearby object compared with more remote reference objects when the nearby object is viewed from two different points in space. In reading several different types of meteorological instruments, errors of parallax are easily introduced if the line of sight were not carefully maintained perpendicular to the reading scale. Parallax errors are of particular concern in reading the level of the liquid of **mercurial barometers** and **liquid-in-glass thermometers** and the position of the pointer on the dial of an **aneroid barometer**.

parameter—A variable coefficient in some model, formula, or other relationship that can be adjusted to apply the general model to particular cases. Typically not an independent variable.

parameterization—The representation, in a dynamic model, of physical effects in terms of admittedly oversimplified **parameters**, rather than realistically requiring such effects to be consequences of the dynamics of the system.

parcel method—A means for testing the condition of static **stability** of the atmosphere, utilizing the behavior of an air parcel that is displaced vertically within an atmospheric layer; the parcel initially is in equilibrium with its environment and undergoes **adiabatic cooling** or **heating** in accordance with either a **dry-adiabatic** or **saturation-adiabatic lapse rate**.

parhelic circle—A **halo** phenomenon consisting of a faint white luminous circle passing through the sun and running parallel to the **horizon** for as much as 360° of **azimuth**; caused by **reflection** of sunlight from airborne ice crystals.

parhelion (pl., **parhelia**)—A **halo** phenomenon representing either of two colored luminous spots that appear at points 22° on both sides of the sun and at the same elevation as the sun; caused by **refraction** of sunlight by ice crystals; also known as *sun dog* or *mock sun*.

partial obscuration—In United States weather observing practice, the designation for **sky cover** when part of the sky (0.1–0.9) is completely hidden by surface-based **obscuring phenomena**.

partial pressure—In a mixture of **ideal gases**, the pressure exerted by one of the gas species.

partial tide (also known as *tidal component*)—One of the harmonic components that compose the tide at any point. The periods of the partial tides are derived from various combinations of the angular velocities of earth, sun, moon, and stars relative to each other.

partly cloudy—(1) In United States climatological practice, the character of a day's weather when the average **cloudiness**, as determined from frequent observations, has been from 0.4 to 0.7 for the 24-hour period. (2) In popular usage, the state of the weather when clouds are conspicuously present, but do not completely dull the day or the sky at any moment. In weather forecast terminology, this term may be used when the expected cloudiness is from about 0.3 to 0.7.

pascal (Pa)—The **SI** derived unit of **pressure** equal to the pressure resulting from a force of 1 newton acting uniformly and perpendicularly over an area of 1 square meter: 1 Pascal = 0.01 hectopascal = 0.01 millibar = 1 newton per square meter. Named for Blaise Pascal (1623–1662), a French physicist who performed early barometry experiments.

Pascal's law—A **hydrostatic** principle that pressure applied to an enclosed **fluid** is transmitted undiminished to every portion of the fluid and to the walls of the containing vessel.

peak discharge—The maximum instantaneous volumetric flow of water (**discharge**) of a given **hydrograph**.

peak wind speed or **peak gust**—In United States weather observing practice, the highest "instantaneous" **wind speed** recorded at a station during a specified time interval, usually the 24-hour observational day.

pearl lightning—Same as **bead lightning**.

pennant—Triangular symbol on a **wind shaft** with a value of 50 knots.

percentage of possible sunshine—Ratio of the actual duration of bright sunshine made by a **sunshine recorder** to the astronomically possible duration of sunshine, as determined by the **sunrise** and **sunset** calculations for that station.

perched aquifer—A water-bearing geological formation with a thin **zone of saturation** located above the main **water table**.

percolation—The downward seepage of liquid water though an unsaturated porous medium such as the soil or a snow layer caused by **gravity**.

perennial stream—A stream that contains water at all times except during extreme **drought.** Contrast with **ephemeral** or **intermittent streams**.

perfect gas—See **ideal gas**.

perigean tide—A tide of increased **tidal range** occurring when the moon is near **perigee**. Contrast with **apogean tide**.

perigee—On the orbit of the moon relative to the earth, the point nearest the earth. Opposite of **apogee**.

periglacial—Of, or pertaining to, the outer perimeter of a glacier, particularly to the fringe areas surrounding the great **continental glaciers** of the geologic **ice ages**. Thus, ''periglacial weathering'' is said to have produced certain characteristic land forms.

periglacial climate—The climate characteristic of the regions immediately bordering the outer perimeter of an **ice cap** or **continental glacier**. The principal climatic feature is the high frequency of very cold and dry winds off the ice area. These regions have been thought to offer ideal conditions for the maintenance of a belt of intense **cyclonic** activity.

perihelion—The point in the earth's orbit nearest the sun. Opposite of **aphelion**. At present, the earth reaches this point (147 million kilometers from the sun) on about 3 January; but the date varies irregularly from year to year (effect of leap year system) and also has a slow secular change of a 1 day advance for every 60 years due to the **precession of equinoxes**.

period—(1) The time interval between passages, at a fixed point, of a given phase of a roughly periodic oscillation or **harmonic** wave; the reciprocal of **frequency**. (2) The time interval for a planet or satellite to complete one revolution on its orbit. (3) See **geologic period**.

period of record—The official, certified length of time during which a specific **weather element** (e.g., temperature, humidity, precipitation) has been observed at a particular place.

permafrost—A layer of soil or bedrock at a variable depth beneath the surface of the earth in which the temperature has been below freezing continuously from a few to several thousands of years. Permafrost exists where the summer heating fails to descend to the base of the layer of frozen ground. A continuous stratum of permafrost is found where the annual mean temperature is below about –5°C (23°F).

permafrost table—The more or less irregular surface in the ground that marks the upper limit of **permafrost**; not to be confused with **frost table**.

permanent anticyclone—An atmospheric **high pressure** system that largely predominates throughout the year in a particular region and therefore appears as a recognizable feature on the mean annual pressure chart.

permanent cyclone—An atmospheric **low pressure** system that largely predominates throughout the year in a particular region and therefore appears as a recognizable feature on the mean annual pressure chart.

permanent wave—A wave (in a **fluid**) moving with no change in **streamline** pattern, and which, therefore, is a **stationary wave** relative to a coordinate system moving with the wave.

permeability—Varying capacity with which water seeps into the ground under the force of **gravity**. It thus expresses the rate of **percolation**.

persistence—In general, the tendency for the occurrence of a specific event to be more probable, at a given time, if that same event has occurred in the immediately preceding time interval. In other words, a continuation of existing conditions.

persistence forecast—In meteorology, a **forecast** that the future weather condition will be the same as the present condition. The persistence forecast is often used as a standard of comparison in measuring the degree of skill of forecasts prepared by other methods.

perturbation—Any departure introduced into an assumed steady state of a system.

Peru Current (also known as the ***Humboldt Current***)—The cold ocean current in the South Pacific flowing northward along the coasts of Chile and Peru from about 40° to 20°S. It is one of the swiftest of ocean currents. The Peru current originates where part of the water that flows toward the east across the Subantarctic Pacific Ocean is deflected toward the north as it approaches South America. The northern limit of the current can be placed a little south of the equator, where the flow turns toward the west, joining the **South Equatorial Current**.

pH scale—A measure of the range of acidity and alkalinity of different substances, extending from 0 to 14. Expressed as the negative logarithm of the hydrogen ion concentration of a solution so that the pH decreases as the hydrogen ion concentration increases. A pH of 7 is neutral; acids have pH values less than 7, and alkaline substances have pH values greater than 7.

phase—(1) The state of aggregation of a substance, for example, solid, liquid, or gas; hence, the "*physical phase.*" (2) A point of reference on a periodic or **harmonic wave**, typically, an angular measure (phase angle) of one cycle. (3) The lunar phase or "*phase of the moon,*" representing one of the recurring shapes of the illuminated portion of the moon apparent to the earthbound observer during a complete **lunation** cycle; for example, full moon, new moon, etc.

phase velocity—Speed of movement of a constant **phase** (2), as represented by the **crests** and **troughs** of a wave. See **group velocity**.

phenology—The systematic study of the chronology of biological and physical events having a periodic or recurring nature in relation to **weather** and **climate**; records are maintained to provide a time series of such *phenological events* as flowering, migration, **freeze over**, and **ice-out.**

photochemical reaction—A chemical reaction involving either the **absorption** or **emission** of **electromagnetic radiation**.

photochemical smog—A visibility-restricting noxious mixture of **aerosols** and gases; the product of **photochemical reactions** involving sunlight, oxides of nitrogen from motor vehicle exhaust, and volatile organic compounds from various anthropogenic and biogenic sources. High levels of ground-level **ozone** result.

photodissociation—The splitting (dissociation) of a molecule into two or more neutral atoms or molecules by the **absorption** of a **photon**. The resulting components may be ionized in the process (**photoionization**). Radiation of ultraviolet or shorter wavelengths are involved. Stratospheric **ozone** is formed by photodissociation processes.

photoionization—The removal of one or more electrons (**ionization**) of an atom or molecule by its collision with a high-energy **photon**, leaving an electrically charged **ion**; this process typically occurs in the **ionosphere**, with very short **ultraviolet** radiation required.

photometeor—Any luminous optical phenomenon occurring in the atmosphere, produced by the **scattering**, **reflection**, **refraction**, **diffraction**, or **dispersion** of visible light from the sun or moon; examples include **coronas**, **haloes**, and **rainbows**.

photometer—An instrument for measuring the intensity of light (**illuminance**) or the relative intensity of a pair of lights.

photon—According to the quantum theory of radiation, the elementary quantity, or "quantum" of **radiant energy**.

photoperiod—The interval during the day when an organism is exposed to sufficient light so as to induce physiological or behavioral effects, to include those associated with seasonal **illumination** changes.

photosphere—The intensely bright portion of the sun visible to the unaided eye. It is a shell several hundred kilometers in thickness marking the boundary between the dense interior gases of the sun and the more diffuse cooler gases in the overlying **chromosphere** and other outer portions of the sun. Features such as **sunspots** and **faculae** are observed on the photosphere.

photosynthesis—The **endothermic** process whereby green plants use sunlight, water, and carbon dioxide in the presence of chlorophyll to manufacture carbohydrates; a by-product is **oxygen**.

photosynthetically active radiation—**Electromagnetic radiation** in the part of the spectrum used by green plants for **photosynthesis**, particularly in two bands centered at 0.45 and 0.65 micrometers.

physical climatology—The major branch of climatology, which deals with the explanation of **climate**, rather than with presentation of it (**climatography**).

physical meteorology—That branch of meteorology dealing with optical, electrical, acoustical, and thermodynamic phenomena of the atmosphere, its chemical composition, the laws of radiation, and the explanation of clouds and precipitation.

pibal—Contraction for **pilot-balloon observation**.

piezometric surface—The imaginary surface that everywhere coincides with the head of the water in the **aquifer**; that is, the **elevation** to which the water would rise in a well connected to the aquifer. In areas of artesian **groundwater**, it is above the land surface.

pileus—An **accessory cloud** that occurs above or attached to the top of a **cumuliform cloud**. See **cap cloud**.

pilot balloon—A small balloon whose ascent at a constant rate is followed by a **theodolite** in order to obtain data for the computation of the speed and direction of winds at various levels in the upper air above the station.

pilot-balloon observation (pibal)—A method of **winds aloft** observation; that is, the determination of **wind speeds** and **directions** in the atmosphere above a station. This is done by reading the **elevation** and **azimuth angles** of a **theodolite** while visually tracking a **pilot balloon**.

pilot report (PIREP)—A report of inflight weather by an aircraft pilot or crew member. A complete coded report includes the following information in this order: location and/or extent of reported weather phenomena; time of observation; description of phenomena; altitude of phenomena; type of aircraft (only with reports of **turbulence** or **icing**).

pingo—A large **frost mound** of more than one-year duration.

PIREP—Acronym for **pilot report**.

Pitot tube—A **pressure-type anemometer** used to measure the relative speed of a **fluid**, often times on an aircraft to measure **wind speed**. It consists of a tube, with one open end held perpendicularly to the fluid flow to detect **dynamic pressure**, which is proportional to the square of the fluid speed. Designed by Henri Pitot (1695–1771), a French physicist.

pixel—A contraction of "*picture element,*" representing the smallest discrete element of an electronically encoded image (having both spatial and spectral components) recorded by a satellite sensor.

Plan Position Indicator (PPI)—Conventional **radar** display on which **range** is indicated by concentric radial distance circles centered on the radar site and direction is given by **azimuth**, as indicated by radial lines emanating from the origin.

Planck's constant—A constant, usually designated h, of dimensions $mass \times length^2 \times time^{-1}$ equal to 6.62196×10^{-32} joules second. It scales the energy of **electromagnetic radiation** of frequency v such that the radiation appears only in quanta nhv, n being an integer. Named for Max Planck (1858–1947), a German physicist and quantum theory pioneer.

Planck's law—An expression derived by Max Planck for the variation of monochromatic **emittance** (emissive power) as a function of wavelength of **blackbody** radiation at a given temperature; the amount of radiation emitted by a blackbody is uniquely determined by its absolute temperature.

planetary albedo—The fraction of incident **solar radiation** that is **scattered** and **re-**

flected back to space by the entire earth–atmosphere system; the currently accepted value for planet earth as determined from satellite measurements is 30%.

planetary boundary layer—The atmospheric layer from the earth's surface up to an altitude of about 1 kilometer in which **wind speed** and direction are affected by frictional interaction with objects on the earth's surface; also called **atmospheric boundary layer** and the **friction layer**. Usually subdivided into the **surface boundary layer** and the **mixed layer** or **Ekman layer**. The **free atmosphere** lies above the planetary boundary layer.

planetary circulation—(1) The system of large-scale disturbances in the **troposphere** when viewed on a hemispheric or **planetary scale**. (2) The mean or time-averaged hemispheric circulation of the atmosphere; in this sense, almost synonymous with **general circulation**.

planetary heat balance—The **equilibrium** that exists on average between the radiation received by the earth–atmosphere system from the sun and that emitted by both the earth and atmosphere to space. The observed long-term constancy of the earth's surface temperature demonstrates that the equilibrium does exist in the mean. On average, regions of the earth nearer the equator than about 35° latitude receive more energy from the sun than they are able to radiate, while latitudes higher than 35° receive less. The excess of heat is carried from low latitudes to higher latitudes by atmospheric and oceanic circulations, and is reradiated there.

planetary scale—The largest space scale, covering essentially the entire planet, and incorporating the **primary** or **planetary circulation** regimes; larger than **synoptic scale**.

planetary wave—Same as **long wave** or **Rossby wave**.

plate tectonics—Concept that the outer 100 kilometers of the solid earth (consisting of the crust plus the rigid upper portion of the mantle) is divided into a number of plates that move relative to one another across the surface of the planet; responsible for **continental drift**.

Pleistocene epoch—The **geologic epoch** of the **Quaternary period** extending from approximately 2 million to 10 800 years ago (the start of the **Holocene** epoch). A large portion of this time interval experienced extensive **Pleistocene glaciation**.

Pleistocene glaciation—The physical extent and/or the geologic results of the most recent ice age, the **Quaternary Ice Age**. It is so named because it occurred in the **Pleistocene epoch** of the **Quaternary period**.

plow wind—A term used in the midwestern United States to describe strong, straight-line winds associated with the **downdrafts** spreading out in advance of the **squall lines** and **thunderstorms**. Resulting damage is usually confined to narrow zones like that caused by **tornadoes**; however, the winds are all in one direction.

plume—Identifiable stream of air with a temperature or composition different from that of its environment. Examples are a smoke plume from a chimney and a buoyant plume rising by convection from heated ground.

pluvial—(1) Pertaining to rain, or more broadly, to precipitation; particularly to an abundant amount thereof. (2) Pertaining to an interval of **geologic time** that was marked by a relatively large amount of precipitation. This term is usually applied to those episodes of heavy rainfall in the lower latitudes associated with the equatorward advance of the glacierization of an **ice age**. Lakes in present-day dry regions expanded. Thus, a pluvial period in low latitudes generally coincides with a **glacial stage** of higher latitudes.

POES (Polar-orbiting Operational Environmental Satellite)—A **polar-orbiting satellite** series operated by the National Oceanic and Atmospheric Administration (NOAA), they are designated ''NOAA satellites.'' Included in this group are the *TIROS-N* satellite series, the third generation of NOAA polar-orbiting environmental spacecraft.

point discharge—A silent, nonluminous, gaseous **electrical discharge** from a pointed conductor maintained at a potential that differs from that of the surrounding gas. Trees and other grounded objects may be sources of point discharge currents.

point of occlusion—On a synoptic chart, the point at which the **warm front**, **cold front**, and **occluded front** intersect; also known as the **triple point** (3).

point source—(1) In air quality studies, a precise spot such as a chimney or exhaust pipe that is a source of air contaminants; contrast with **area source** and **line source**. (2) Source of radiation that is small in size with respect to its distance from an irradiated target. The **flux** per unit area of the radiation field varies inversely as the square of the distance from the source. See **inverse square relationship**.

point target—In **radar meteorology**, an object that returns a **target signal** by reflection from a relatively simple discrete **target** surface. Such targets are ships, aircraft, projectiles, missiles, buildings, etc.

polar air—An **airmass** type whose characteristics are developed over high latitudes, especially within the subpolar highs. Contrast with **tropical air**. Continental polar air (*cP*) has low surface temperature, low humidity, and especially in its **source regions**, has great stability in lower layers. Maritime polar air (*mP*) initially possesses similar properties to those of continental polar air, but in passing over warmer water it becomes unstable with a higher water vapor content.

polar anticyclone—Same as **arctic high**.

polar climate—A **climatic zone** located in the polar latitudes, marked by conditions too harsh to support vegetation. See the **Frigid Zone**.

polar cyclone—Same as **polar vortex**.

polar easterlies—The rather shallow and diffuse body of easterly winds (winds from east) located poleward of the **subpolar low-pressure belt**. On a mean wind map of the Northern Hemisphere, these easterlies exist to an appreciable extent only north of the **Aleutian low** and **Icelandic low**.

polar front—The semipermanent, semicontinuous boundary (**front**) found in midlatitudes separating air masses of **tropical** and **polar** origin. This is the major front in terms of airmass contrast and susceptibility to **cyclonic** disturbance.

polar front theory—A theory originated by the Scandinavian school of meteorologists in the early twentieth century whereby a **polar front**, separating **air masses** of polar and tropical origin, gives rise to **cyclonic** disturbances that intensify and travel along the front, passing through various phases of its life cycle. See **Norwegian cyclone model**.

polar glacier—A **glacier** whose **accumulation zone** is covered by **firn**, and whose subsurface temperatures are below the pressure-melting point throughout the year.

polar high—An **anticyclone** that corresponds to a dome of continental polar air (see **polar air**); same as **arctic high**.

polar ice—(1) The thickest form of **sea ice**, one to several years old and sometimes more than three meters thick. (2) The **pack ice** of the central Arctic Ocean.

polar jet stream—A zone of unusually strong winds located between the midlatitude **tropopause** and the polar tropopause. It is associated with the latitudes of greatest north to south **temperature gradient**, in particular, over the **polar front** and the tracks of midlatitude **cyclones**.

polar low—A relatively small (several hundred kilometers) migratory **cyclone** in **polar air**, poleward of **frontal zones** during the cold season. Generally formed along ice/land boundaries and relatively warm water.

polar orbit—An orbit with an orbital **inclination** of near 90° where the satellite ground track will cross both polar regions once during each orbit. The term is used to describe the near-polar orbits of spacecraft such as the NOAA/TIROS and Landsat satellites.

polar outbreak—The movement of a cold **air mass** from its **source region**; almost invariably applied to a vigorous equatorward thrust of cold **polar air**, a rapid equatorward movement of the **polar front**.

polar trough—In tropical meteorology, a wave **trough** in the **circumpolar westerlies** having sufficient amplitude to reach the Tropics in the upper air. At the surface it is reflected as a trough in the **tropical easterlies**, but at moderate elevations it is characterized by westerly winds.

polar vortex (or *circumpolar vortex*)—The large-scale **cyclonic** circulation in the middle and upper **troposphere** centered generally in the polar regions. Specifically, the vortex in the Northern Hemisphere has two centers in the mean, one near Baffin Island and another over northeast Siberia. The associated cyclonic wind system comprises the westerlies of middle latitudes.

polarity—The sign of the electric charge associated with a given object, as an electrode or an **ion**.

polarization—The state of **electromagnetic radiation** when transverse vibrations take place in some regular manner, for example, all in one plane, in a circle, in an ellipse, or in some other definite curve. Radiation may become polarized because of the nature of its emitting source, as is the case with many types of radar antennae, or because of some processes to which it is subjected after leaving its source, as the polarized light resulting from the **scattering** of solar radiation as it passes through the earth's atmosphere.

polar-orbiting satellite—A satellite whose orbital plane passes near the North and South Poles. The earth rotates under the plane of the satellite.

pole—(1) For any circle on the surface of a sphere, the point of intersection of the sphere's surface and the normal line through the center of the circle. (2) A point at which the axis of rotation passes through the surface. The North and South geographic Poles are rotation poles and the poles, as defined from (1) above, of the equator or of any other latitude circle. (2) A point of concentration of **electric charge** or **magnetic force**. See, for example, one of the poles of the **dipolar** water molecule or the **geomagnetic poles** of the earth.

poleward heat transport—A **meridional flow** of **sensible** and **latent heat** from tropical to middle and high latitudes in response to latitudinal imbalances in radiational heating and cooling; brought about by airmass exchange, storms, and surface **ocean currents**.

pollen analysis—Study of the distribution of pollen grains of various species contained in surface-layer deposits, especially in peat bogs, from which indication about changes in climate may be inferred.

pollutant—Any natural or anthropogenic substance that impairs the suitability of the air or water for a considered purpose. Compare with **contaminant**.

Pollutant Standards Index (PSI)—Standardized index used in the United States to report the level of **air pollution** in ambient air; based on the **criteria air pollutant** that exhibits the highest concentration when compared to its **National Ambient Air Quality Standard**.

pollution—Any alteration of the natural environment by either natural or anthropogenic sources creating a condition that is harmful to living organisms. Examples include **atmospheric pollution** and water pollution.

polynya—A large area of open water surrounded by **fast ice** or **sea ice**. This open water area remains relatively constant and usually has an oblong form.

porosity—The ratio of the total amount of void or pore space in a material to the total bulk volume of that material. A measure of the capacity of soil or rock to hold water.

potential energy—The **energy** that a body possesses as a consequence of its position in the field of **gravity**; numerically equal to the **work** required to bring the body from an arbitrary standard level, usually taken as **mean sea level**, to its given position.

potential evaporation—The quantity of water that would be emitted into the atmosphere as a **vapor** from a surface of pure water in the existing conditions. The evaporation from a **pan evaporimeter** provides a good estimate.

potential evapotranspiration—Generally, the amount of water vapor that, if available, would be removed from a given well-watered land area by **evapotranspiration**; expressed in units of water depth. Usually this amount is the maximum amount for the particular climate and vegetation cover. Contrast with **actual evapotranspiration**.

potential temperature—(1) The temperature a dry **air parcel** would have if brought dry adiabatically from its initial state to the **standard pressure** of 1000 hectopascals (millibars). (2) In oceanography, the temperature that a water sample would attain if raised adiabatically to the sea surface. For the deepest points of the ocean, which are just over 10 000 meters (32 800 feet), the **adiabatic cooling** would be less than 1.5°C (2.7°F).

powder snow—A skiing term for a loose cover of fresh, dry snow that has not been compacted in any way.

power—The time rate, often expressed in **watts**, at which **energy** is fed to or taken from a device; generally analogous to **flux**. In **radar meteorology**, it usually refers to the time rate at which energy is radiated from or received at the antenna.

PPI—Acronym for **Plan Position Indicator**.

ppm—Abbreviation for parts per million, a convenient measure of concentration.

practical salinity—A measure of the concentration of dissolved salts in **seawater** (**salinity**) based upon electrical conductivity. Compare with **absolute salinity**. Values of salinity are expressed in **practical salinity units (psu)**.

practical salinity unit (psu)—The dimensionless unit used to describe **practical salinity**. The average salinity of seawater is 35 psu.

prairie—A flat or gently undulating plain that is grassy and generally treeless, except for along river banks; specifically, such an area in southern Canada and the northern and central United States where it extends from the foothills of the Rocky Mountains eastward to about 88°W. A large portion of region is semiarid, with high summer temperatures. See **steppe**.

precession of the equinoxes—The wobble of the earth's axis of rotation having a period of about 21 500 years because of luni-solar gravitational effects; this effect is seen as a slow rotation of constellations along the **zodiac**. A contributor to the **Milankovitch cycles**, affecting seasonal length, the seasonal solar radiation receipt, and the timing of **perihelion** passage, with an advancement of the **perihelion** date (presently 3 January) by approximately one day every 60 years.

precipitable water—The total atmospheric water vapor contained in a vertical column of unit cross-sectional area extending between any two specified levels, commonly expressed in terms of the height to which that water substance would stand if completely condensed and collected in a vessel of the same unit cross section. The *total precipitable water* is that contained in a column of unit cross section extending all of the way from the earth's surface to the "top" of the atmosphere.

precipitation—(1) Any or all of the forms of water particles, whether liquid or solid, that fall from clouds and reach the ground. It is a major class of **hydrometeor** but is distinguished from **cloud**, **fog**, **dew**, **rime**, **frost**, etc., in that it must "fall"; and is distinguished from cloud and **virga** in that it must reach the ground. Precipitation includes *liquid precipitation* (**drizzle**, **rain**), *freezing precipitation* (**freezing drizzle**, **freezing rain**), and *frozen precipitation* (**snow pellets**, **snow grains**, **ice crystals**, **ice pellets**, **hail**). (2) The amount, usually expressed in units of liquid water depth (e.g., inches), of the water substance that has fallen on a horizontal surface at a given point over a specified time interval. As this is usually measured in a fixed **rain gauge**, small amounts of dew, frost, rime, etc., may be included in the total.

precipitation attenuation—The loss of **microwave** radar energy due to passage through a volume of the atmosphere containing precipitation. Part of the energy is lost by **scattering** and part by **absorption**.

precipitation echo—A type of **radar echo** returned by precipitation particles.

precipitation effectiveness—The efficacy of precipitation in plant growth.

precipitation gauge—General term for any device that measures the amount of precipitation; principally, a **rain gauge** or **snow gauge**.

precipitation intensity—The time rate of precipitation, usually expressed in millimeters (inches) per hour, representing the rate of precipitation collected in a **recording rain gauge** or estimated by **radar reflectivity** or by a reduction in **visibility**.

precipitation shadow—Same as **rain shadow**.

prediction—The act of making a **forecast** of a future occurrence, such as a weather event; or, the forecast itself.

pressure—In meteorology, commonly used for **atmospheric pressure**.

pressure altimeter—An **aneroid barometer** calibrated to convert **atmospheric pressure** into **altitude**. Altimeters use **standard atmosphere** pressure–height relations in converting pressure into altitude.

pressure altitude—The height above mean sea level (**altitude**) corresponding to a given value of atmospheric pressure according to the **standard atmosphere**; this would be the **indicated altitude** of a **pressure altimeter** with an **altimeter setting** of 29.92 inches or 1013.25 hectopascals (millibars).

pressure center—On a **synoptic chart** (or on a chart of mean atmospheric pressure), a point of local minimum or maximum pressure, or **central pressure** of the system; the center of a **low** or **high**. A pressure center is also a center of **cyclonic** or **anticyclonic** circulation.

pressure force—The force due to differences of **pressure** within a **fluid** mass; strictly the **pressure gradient force**.

pressure gradient—The rate of decrease (gradient) of **pressure** in space at a fixed time. The term is sometimes loosely used to denote simply the magnitude of the **gradient** of the pressure field. The three-dimensional pressure gradient **vector** can be resolved into *vertical* and *horizontal* components.

pressure gradient force—A three-dimensional force vector operating in the atmosphere that accelerates **air parcels** away from regions of high pressure and toward regions of low pressure in response to an air **pressure gradient**. Usually resolved into the *vertical* and *horizontal* pressure gradient force components.

pressure head—The height of a column of nonflowing (static) water that can be supported by the **static pressure** at a point.

pressure ice—Floating **sea ice** (or river or lake ice) that has been deformed, altered, or forced upward in **pressure ridges** by the lateral stresses of any combination of wind, water currents, tides, waves, and surf.

pressure jump—A sudden, sharp increase in **atmospheric pressure**, typically occurring along an **active front** and preceding a storm. See **hydraulic jump**.

pressure pattern—In meteorology, the general geometric characteristics of the atmospheric pressure distribution as revealed by isobars on a **constant-height chart**; usually applied to **synoptic-scale** features of a surface chart.

pressure ridge—A discernible rise or ridge, up to 30 meters (90 feet) high and sometimes several kilometers (several miles) long, in **pressure ice**.

pressure system—An individual synoptic-scale feature of atmospheric circulation, commonly used to denote either a **high** or a **low**; less frequently a **ridge** or a **trough**.

pressure tendency—The character and amount of atmospheric pressure change for a three-hour or other specified period ending at the time of observation.

pressure-type anemometer—Any instrument that uses pressure to determine the speed of a **fluid**. Examples include the pressure plate anemometer and the **Pitot tube**.

prevailing visibility—In United States weather observing practice, the **greatest horizontal visibility** that is equaled or surpassed through at least one-half of the horizon circle.

prevailing wind direction—The **wind direction** most frequently observed during a given period.

primary air pollutant—A substance that is an **air pollutant** immediately upon release to the atmosphere; examples include carbon monoxide, sulfur dioxide, and certain hydrocarbons. Contrast with **secondary air pollutant**.

primary circulation—The prevailing fundamental atmospheric circulation on a **planetary scale** that must exist in response to (*a*) radiation differences with latitude, (*b*) the rotation of the earth, and (*c*) the particular distribution of land and oceans; and which is required from the viewpoint of **conservation of energy**. Contrast with **secondary** and **tertiary circulation**.

primary rainbow—The most common of the principal **rainbow** phenomena, which appears as an arc of angular radius of about 42° about the observer's **antisolar point**. Its spectral color sequence is from red on the outside to violet on the inner edge. The primary rainbow results from a single internal **reflection** within raindrops plus two **refractions**, one on entering and one on leaving the drop; **dispersion** of **polychromatic** sunlight into component colors results from the wavelength-dependency of **refraction** of light waves passing through water droplets. See **secondary rainbow**.

Prime Meridian—The 0° longitude line (which passes through Greenwich, England) from which **longitude** is measured over 180° eastward and 180° westward, to the meridian essentially coinciding with the International Date Line.

primitive equations—The basic dynamic **equations of motion**, supplemented with the thermodynamic equations, employed in numerical models without simplification or approximation.

probability forecast—A forecast containing a statement of the likelihood of the occurrence of a specific event during a specific time interval at a particular place.

probability of precipitation (**PoP**)—(1) A forecast guidance product defined as the likelihood of occurrence of a measurable amount of **liquid precipitation** (or **water equivalent** of **frozen precipitation**) during a specified time interval (e.g., 12 hours) at any given point in the forecast area; expressed as a percentage. Compare with **quantitative precipitation forecast (QPF)**. (2) In a general public forecast, the forecast statement of precipitation is described in terms of the likelihood of occurrence in terms of a percentage. The following terms are typically used: **slight chance**, **chance**, and **likely**.

process lapse rate—The theoretical rate at which an **air parcel** cools with height (**lapse rate**) as the parcel is lifted and undergoes either a **dry-** or **saturation-adiabatic process**, depending upon whether the parcel were unsaturated or saturated; hence, the process lapse rate refers either to **a dry-adiabatic lapse rate** or a **saturation-adiabatic lapse rate**. Contrast with **environmental lapse rate**.

profile—In meteorology, a graph of the value of a **scalar** quantity versus a horizontal, vertical, or timescale. It usually refers to a vertical representation, such as the temperature profile.

profiler—See **wind profiler**.

prog—Common contraction for **prognostic chart**.

prognosis—Forecaster's judgment, based on an **analysis** or the forecast guidance output of a **numerical weather prediction** model, of the future development of a weather situation.

prognostic chart—A chart showing, principally, the expected **pressure pattern** (or **height pattern**) of a given **synoptic chart** at a specified future time. Anticipated positions of surface **fronts** are also included, and the forecast values of other **weather elements** may be superimposed. Usually, these charts are the graphical output of a **numerical weather prediction** model. Distinguish from a **diagnostic chart**.

propeller anemometer—A **rotation anemometer** with a radiating array of propeller-shaped blades mounted on a horizontally mounted shaft; the device is allowed to point into the wind and the rotation rate of the propellers is proportional to the **wind speed**.

pseudoadiabat—On a **thermodynamic diagram**, a line representing the change in temperature of a **saturated air parcel** as it undergoes a **pseudoadiabatic expansion**.

pseudoadiabatic expansion—A **saturation-adiabatic process** in which the condensed water substance is removed from the system. Meteorologically, this process corresponds to rising air from which the moisture is precipitating. Descent of air so lifted becomes, by definition, a **dry-adiabatic process**.

pseudoadiabatic lapse rate—The rate of temperature change with height that a **saturated air parcel** experiences in a **pseudoadiabatic expansion** process; a **process lapse rate**. For practical purposes in meteorology, same as **saturation-adiabatic lapse rate**.

psychrometer—An instrument used to measure the water vapor content of the air; a type of **hygrometer**, such as a **sling** or **aspirated psychrometer**. It consists of two **liquid-in-glass thermometers**, one of which (*dry-bulb*) measures the ambient air temperature. The bulb of the other thermometer (*wet-bulb*) is covered with a muslin sock that is saturated with distilled water prior to an observation. When ventilated, the instrument indicates the **dry-bulb** and **wet-bulb** temperatures. By referring to special **psychrometric tables**, the **relative humidity** and/or **dewpoint** can be determined.

psychrometric table—A tabulation prepared from a semiempirical formula used to obtain the dewpoint, vapor pressure, or the relative humidity of the air from the observed **dry-bulb** and **wet-bulb** temperatures obtained by a **psychrometer**.

pulse—In **radar**, a single short-duration transmission (or ''burst'') of microwave energy as transmitted by **pulse radars**.

pulse length—The linear distance occupied by a **pulse** of transmitted radio energy in space.

pulse radar—A type of **radar** that emits a timed or pulsed burst of microwave energy and during the intermediate quiet time awaits the return of the reflected signal to determine the distance between the **target** and the radar unit. The majority of weather radar units are pulse radars.

pulse repetition frequency (PRF)—The number of (frequency) energy **pulses** transmitted per unit time by a **pulse radar**. The PRF determines the range of pulse radars, the lower the PRF, the greater the **range**.

purple light—The faint purple glow observed on clear days over a large region of the western sky after **sunset** and over the eastern sky before **sunrise**; typically the sun is between 3° and 6° below the horizon; a **twilight phenomenon**.

pyranometer—General name for the class of radiation sensors that measure the combined intensity of incoming **direct solar radiation** and **diffuse sky radiation**. The pyranometer is mounted so that it views the entire sky hemisphere.

Q

quantitative precipitation forecast (QPF)—Forecast of precipitation amount, specified in depth units (e.g., inches). Compare with **probability of precipitation (PoP)** forecasts.

quantum theory—The theory first stated by Max Planck (1858–1947), a German physicist, before the Physical Society of Berlin on 14 December 1900 that all **electromagnetic radiation** is emitted and absorbed in "quanta," each of magnitude $h\nu$, h being **Planck's constant** and ν the **frequency** of the radiation.

quasi-biennial oscillation (QBO)—See **biennial wind oscillation**.

quasigeostrophic approximation—A form of the **equations of motion** that selectively includes the **geostrophic wind** rather than the actual wind.

quasi-hydrostatic approximation—The use of the **hydrostatic equation** as the vertical **equation of motion**, thus implying that the vertical **accelerations** are small without constraining them to be zero.

quasi-stationary front—A **front** that is stationary or nearly so. See **stationary front**. Conventionally, a front that is moving at a speed less than about five knots is generally considered to be quasi-stationary.

Quaternary Ice Age—Any of the **glaciations** dominating the **Quaternary Period**, featuring numerous large-scale advances and recessions of **ice sheets** over portions of North America and northwestern Europe.

Quaternary Period—The **geologic period** encompassing the last 2 million years and containing the **Pleistocene** and **Holocene** epochs. Notable during this period have been the **Quaternary Ice Age** (or **Pleistocene glaciation**) during the Pleistocene and the present **interglacial** of the Holocene.

R

radar (*ra*dio *d*etection *a*nd *r*anging)—In meteorology, an electronic instrument that broadcasts and receives **microwave** signals back from **targets** for the purpose of determining the location, height, movement, and intensity of precipitation areas; based on the property of precipitation particles (rain, hail, snow) to reflect and scatter microwaves. Some **weather radar** units are **Doppler radars**, which operate either in a *reflectivity* mode or Doppler *velocity* mode, with the latter based on the **Doppler effect**.

radar altitude—The height above ground level (**altitude**) determined by a radar-type **radio altimeter**; essentially, the radar altitude is the same as the **absolute altitude**.

radar echo—That portion of the pulsed beam of **microwave** energy that is reflected back to the receiver after the beam encounters an obstruction or **target** in the atmosphere; ultimately displayed on a **radar screen**.

radar meteorology—(1) Study of the **scattering** of radar waves by various atmospheric phenomena. (2) Use of radar for making weather observations and short-term forecasts.

radar reflectivity—In general, the measure of the efficiency of a radar **target** in intercepting and returning energy from the outgoing radar beam. It depends upon the size, shape, aspects, and the dielectric properties at the surface of the target. It includes the effects not only of **reflection** but also **scattering** and **diffraction**. In particular, the radar reflectivity of a group of **hydrometeors** is dependent upon such factors as (*a*) the drop size distribution; (*b*) the number of particles per unit volume; (*c*) the physical state of the hydrometeor (ice or water); (*d*) the shape or shapes of the individual elements of the group; and (*e*) if asymmetrical, their aspect with respect to the radar. Hence, radar reflectivity can provide an estimate of **rainfall intensity**. See **VIP level**.

radar report (**RAREP**)—The encoded and transmitted report of a radar meteorological observation. These reports usually give the **azimuth**, distance, **altitude**, **intensity**, shape and movement, and other characteristics of **precipitation echoes** observed by the radar.

radar screen—The cathode ray tube in a **radar** unit on which radar **echoes** from **targets** detected by the radar are visually displayed. These displays can be set to either **Plan Position Indicator** (**PPI**) or **Range-Height Indicator** (**RHI**) mode.

radar volume—The volume in space that is irradiated by a given radar unit; depends upon the **pulse length** of the unit.

radial inflow—Component of **wind velocity** directed inward toward the center of an atmospheric system; diametrically opposed to *radial outflow*.

radial velocity—That component of motion toward or away from a point. This terminology usually refers to the component of motion that is parallel to the beam of a **Doppler radar** unit and therefore can be detected.

radian—An **SI** supplementary unit of angular measurement. Defined as the plane angle between two radii of a circle that subtend an arc on the circumference equal in length to the radius of that circle. One radian is approximately 57.3°. See **degree** (2).

radiance—In **radiometry**, a measure of the intrinsic **radiant energy** flux intensity emitted by a **radiator** in a given direction, expressed in units of energy per unit time per unit solid angle; a counterpart of photometric **luminance**. Compare with **emittance**.

radiance temperature—(1) The temperature of a **blackbody** radiating the same amount of **radiant energy** per unit area at the wavelengths under consideration as the observed body. (2) The apparent temperature of a nonblackbody determined by measurement with a **radiometer**.

radiant energy—The energy of any type of **electromagnetic radiation**.

radiation—(1) The process by which **electromagnetic radiation** is emitted or propagated through space by virtue of joint undulatory variations in the electric and magnetic fields. See discussion in **heat transfer**. (2) The process by which energy is propagated through any medium by virtue of the wave motion of that medium, as in the propagation of **sound waves** through the atmosphere or ocean waves along the water surface. (3) Same as **radiant energy** and **electromagnetic radiation**.

radiation belts—Same as **Van Allen radiation belts**.

radiation budget—An accounting procedure where the **net radiation** in a system is specified from inputs and outputs of **solar radiation** and **terrestrial radiation**.

radiation fog—A major type of **fog**, produced over a land area when **radiational cooling**, especially at night, reduces the air temperature to or below its **dewpoint**. See **ground fog**.

radiation frost—The occurrence of **frost** as a result of nocturnal **radiational cooling** when air temperatures fall below 0°C (32°F). This type of frost is often responsible for causing damage to the agricultural crops especially in early autumn when clear nights, low atmospheric humidities, near calm wind conditions, and longer lengths of night increase radiative heat loss from the surface. Contrast with **advection frost**.

radiation inversion—A **temperature inversion** produced by nocturnal **radiational cooling** of the earth's surface, a snow or ice surface, or the upper part of a cloud layer, etc.

radiation laws—The four physical laws which, together, fundamentally describe the behavior of **blackbody radiation**. (*a*) **Kirchhoff's law** is essentially a thermodynamic relationship between **emission** and **absorption** of any given wavelength at a given temperature. (*b*) **Planck's law** describes the variation of intensity of blackbody radiation at a given temperature, as a function of wavelength. (*c*) The **Stefan–Boltzmann law** relates the time rate of radiant energy emission from a blackbody to its absolute (Kelvin) temperature. (*d*) **Wien's displacement law** relates the wavelength of maximum intensity emitted by a blackbody to its absolute temperature.

radiation shield—A device used on certain types of instruments to prevent unwanted **radiation** from biasing the measurement of a quantity.

radiation thermometer (also known as *radiation pyrometer*)—A thermal sensor instrument (**thermometer**) that measures the thermal **radiation** emitted from a source and displays this flux as the radiation temperature of the source; this instrument can be used remotely. See **radiometer**.

radiational cooling—In meteorology, the cooling of an object or a surface, especially the earth's surface and adjacent air, accomplished (mainly at night) whenever the earth's surface suffers a net loss of **heat** due to **terrestrial radiation** being emitted upward from the surface toward space.

radiative equilibrium temperature—The temperature of an object when its rate of **emission** of radiation equals its rate of **absorption** of radiation.

radiator—Any source of **radiant energy**, especially **electromagnetic radiation**.

radio acoustic sounding system (RASS)—The use of upwardly directed **sound waves** with a **wind profiler** to determine the temperature profile.

radio altimeter—An electric instrument used for determining the **radar** (or **absolute**) **altitude** of an aircraft above the earth's surface; radio waves emitted by the instrument and reflected back from the surface are used to determine the altitude directly.

radio energy—All **electromagnetic radiation** of greater wavelength than **infrared radiation** or essentially 1000 micrometers; this region of the **electromagnetic spectrum** includes **microwave** radiation and all radiation in television and conventional radio signals.

radioactive dating—A method of age determination based on the property of radioactive decay of naturally occurring isotopes, an example is **radiocarbon dating**.

radioactivity—(1) The property, possessed by some of the heavy elements, of spontaneously disintegrating into simpler elements accompanied by emission of **alpha particles**, **beta particles**, or **gamma rays**. (2) The number of spontaneous disintegrations per unit mass and per unit time of a given unstable (radioactive) element. The **curie** is a unit of radioactivity.

radiocarbon dating—A **radioactive dating** method for determining the age of organic material based on the natural radioactive decay of carbon-14, an isotope with a half-life of 5570 years.

radiometer—An instrument for measuring **radiant energy**; can also be used to estimate an object's temperature remotely using **radiation laws**. See **radiation thermometer**. The instrument is used onboard satellites to produce images in various wavelength bands.

radiometry—A scientific discipline involved with the study of **electromagnetic radiation**, to include all aspects of its measurement. The primary region of interest is the portion of **the electromagnetic spectrum** containing **ultraviolet**, **visible**, and **infrared radiation**. *Photometry* is an allied subdiscipline specializing in the study of **light**.

radiosonde—A small balloon-borne instrument package equipped with a radio transmitter that measures vertical **profiles** (**soundings**) of temperature, pressure, and humidity in the atmosphere. See **rawinsonde**.

radiosonde balloon—A helium-filled balloon used to carry a **radiosonde** aloft. They are considerably larger than **pilot balloons** or ceiling balloons.

radiosonde observation (**RAOB**)—An evaluation of the temperature, humidity, and pressure aloft, received via radio signals from a balloon-borne **radiosonde**; the height (specifically, **geopotential height**) of each **mandatory** and **significant pressure level** of the observation is computed from these data. See **rawinsonde** for determination of **winds aloft**.

radome—A spherical structure composed of material transparent to **microwave** energy, used to cover the antenna assembly of a radar to protect it from wind and weather.

radon (**Rn**)—A colorless and odorless radioactive gaseous element formed from the decay of radium in soil or rock; a minute constituent of air near the ground. An indoor pollutant considered to be carcinogenic.

rain—A type of **liquid precipitation** in the form of liquid water drops with diameters greater than 0.5 millimeters (0.02 inches), or, if widely scattered, the drops may be smaller. The only other form of liquid precipitation, **drizzle**, is to be distinguished from rain in that drizzle drops are generally less than 0.5 millimeter (0.02 inches) in diameter, are more numerous, and reduce **visibility** much more than does light rain.

rain gauge—An instrument used to collect and measure the accumulated depth of rainwater that has fallen on a unit area during a specified time interval at a given point; may be either **recording** or **nonrecording gauges**; some types measure **rainfall intensity**.

rain gauge shield (or *wind shield*)—A protective attachment placed around the orifice of a **rain gauge** to eliminate the influence of wind eddies on the **catch**. See **Alter shield**.

rain intensity gauge—An instrument that collects, measures, and automatically records the instantaneous rate at which rain is falling on a given surface. Examples include **tipping bucket rain gauge** and **weighing rain gauge**.

rainbow—A **photometeor** consisting of any one of a family of circular arcs consisting of concentric colored bands, arranged from red on the outside to blue on the inside, which may be seen on a ''sheet'' of water drops (rain, fog, spray) illuminated by direct beams of sunlight. The common center of the arcs is on the line connecting the observer's eyes with the existing light (sun, moon, artificial light) looking in the direction of the observer's shadow. See **primary** and **secondary rainbow**.

raindrop—A nearly spherical drop of liquid water with a diameter greater than 0.5 millimeters (0.02 inches) falling through the atmosphere. In careful usage, falling drops with diameters lying in the interval 0.2–0.5 millimeters (0.008–0.02 inches) are called **drizzle** drops rather than raindrops, but this distinction is frequently overlooked and all drops with diameters in excess of 0.2 millimeters (0.008 inches) are called raindrops.

rainfall—The amount of precipitation of any type; usually taken as that amount measured by means of a **rain gauge** (thus, a small, varying amount of direct **condensation** in the form of **dew** is included).

rainfall intensity—The rate at which rain falls and is collected, expressed in units of depth per unit time interval (such as inches per hour). May be measured directly by a **rain intensity gauge** or estimated from **radar reflectivity**.

rainforest—Generally, a forest that grows in a region of heavy annual precipitation, such as in a **humid climate**. Two types are distinguished: (*a*) the **tropical rainforest** (often simply called the ''rainforest''); and (*b*) the **temperate rainforest**.

rain shadow—The region, on the **lee side** of a mountain or mountain range, where the precipitation is noticeably less than on the **windward** side. A good example of a rain shadow in the United States is the arid region east of the Sierra Nevada, since the prevailing westerly winds from over the Pacific Ocean deposit most of their moisture on the western slopes of the range as a result of **orographic precipitation**.

rainy season—In certain types of climate, an annually recurring period of one or more months during which precipitation is a maximum for that region. Opposite of **dry season**.

range—(1) The difference between the maximum and minimum of a given set of numbers; in a periodic process it is twice the amplitude—that is, the height of the wave. See **tidal range**. (2) The distance between two objects, usually an observation point and an object under observation. (3) A maximum distance attributable to some process, as in **visual range**, radar range, or the ''range'' of an aircraft.

range marker—(1) The index marks displayed on radar indicators to establish the scale or facilitate determination of the range of a **target** from the radar. (2) Suitable landmarks of known distance for determination of **horizontal visibility**.

range-height indicator scope (RHI)—A conventional radar display, in the form of a vertical **cross section**, of the meteorological **targets** encountered in a vertical plane having a specific azimuth. This mode permits the determination of the height of the precipitation shaft.

RAOB—Contraction for **radiosonde observation**.

rare gas—Same as **inert gas**.

rating curve—For a given point on a stream, a graph of **discharge** versus **stage**. For streams of steep slope (several feet or more per mile), the curve is usually a simple-valued function (which may change with time). For large rivers of slight slope, the rating curves are functions of the slope of the water surface. Same as *stage-discharge relation*.

raw—Colloquially descriptive of uncomfortably cold weather, usually meaning cold and damp, but sometimes cold and windy.

rawinsonde—A method of **upper-air observation** consisting of an evaluation of the **wind speed** and **direction**, as well as temperature, pressure, and humidity aloft by means of a balloon-borne **radiosonde** (instrument package) tracked utilizing position change as determined by directional radio techniques.

ray—An elemental path of radiated energy; or the energy following this path.

Rayleigh scattering—Any **scattering** process produced by spherical particles whose radii are smaller than about one-tenth the wavelength of the scattered radiation; explains the blue of the daytime sky. Named for Lord Rayleigh (John W. Strutt, 1842–1919), an English physicist and mathematician who first described this effect. Contrast with **Mie scatter**.

reach—A definite portion of a stream **channel**, commonly taken between two **gauging stations**, but may be taken between any two specified endpoints, such as defined cross sections.

real time—As it happens; usually describing the rapid acquisition, processing, and transmission of data.

recession—(1) The decrease in **streamflow**, at a point along a stream **channel**, following the passage of a **crest**. (2) That portion of a **hydrograph** identified as a **falling limb** that shows the rate of decrease of **stage** or **discharge** following the passage of a crest. Opposite of **rising limb**.

recession curve—A smoothed composite of the **recessions** of several observed **hydrographs**, drawn to represent the characteristic time graph of decreasing total **runoff** for a **drainage area** after passage of a **peak flow**.

recharge—A process where the water in an **aquifer** is replenished by the addition of water from outside the **zone of saturation**.

recharge area—A region where absorbed water eventually reaches an **aquifer** in the **zone of saturation** through surface **infiltration**.

recombination—In **atmospheric electricity**, the **exothermic** process by which a positive and negative **ion** join to form a neutral species. This process occurs in the **return streamer** of a **lightning discharge**. Contrast with **ionization**.

record observation—A type of **aviation weather observation** that is encoded and transmitted; the most complete summary of weather conditions for all such observations and usually taken at regularly specified and equal intervals (hourly, usually on the hour). Contrast with **special observation**.

recording rain gauge—An instrument that automatically records the amount of precipitation collected, as a function of time; used to estimate **rainfall intensity**. The **tipping bucket rain gauge** and **weighing rain gauge** are two examples.

recurrence interval (also known as *return period*)—A statistical parameter used in frequency analysis as a measure of the average time interval between the occurrence of a given quantity and that of an equal or greater quantity; used in hydrology for providing an average time interval between floods of a given magnitude.

red tide—A growth of dinoflagellates (single-celled plantlike animals) in surface waters in such quantities as to color the sea red and kill fish.

reduced pressure—Calculation of **atmospheric pressure** at a standard level from the pressure measured at another level (station pressure or actual pressure) by taking into account, according to theory, the weight of a column of air between the two levels. Pressures displayed on surface weather maps, unless noted otherwise, have been reduced to sea level.

reduction to sea level—An adjustment applied to surface air pressure readings or **station pressure** in order to obtain a value of air pressure that would have been obtained at sea level, thereby eliminating the influence of **station elevation** upon air pressure readings. Standardized "reduction procedures" must be applied to the **station pressure** to obtain a sea level corrected pressure.

reference frame—Same as **coordinate system**.

reflectance—See **albedo**.

reflection—The process whereby a surface turns back a portion of the incident radiation into the medium through which the radiation approached the surface; this process differs from **scattering** in that more radiation is directed backward. **Specular** and **diffuse reflection** result from the types of reflecting surfaces; both types of reflection contribute to atmospheric optical phenomena.

reflectivity—A measure of the fraction of **radiation** reflected by a given surface; defined as the ratio of the radiant energy reflected to the total that is incident upon that surface; usually refers to radiation in a small wavelength band. Compare with **albedo**. Also refers to **radar reflectivity**.

refraction—The process in which the direction of energy propagation is changed as the result of a change in density within the propagating medium, or as the energy passes through the interface representing a density discontinuity between two media. This process contributes to numerous optical phenomena in the atmosphere. Since refraction is wavelength-dependent, the refraction of polychromatic light results in **dispersion**.

regional forecast—In general, a weather **forecast** for a specified geographic region. Contrast with **local forecast**.

relative humidity—The (dimensionless) ratio of the actual **vapor pressure** of the air to the **saturation vapor pressure** for the ambient air temperature. The corresponding ratios of **specific humidity** or of **mixing ratio** give approximations of sufficient accuracy for most purposes in meteorology.

relative vorticity—A measure of the rotation of **fluid** (**vorticity**) as measured in a system of coordinates fixed on the earth's surface. Usually, only the vertical component of the vorticity is meant. Positive values of the vertical component of relative vorticity indicate **cyclonic** spin (in same direction as the earth's rotation) relative to the earth's surface, while negative values identify **anticyclonic** spin. Distinguish from **absolute vorticity**.

remote sensing—The technology of measurement or acquiring data and information about an object or phenomena by a device that is not in physical contact with it; examples in meteorology include radar and sensors onboard satellites. Contrast with **in situ**.

reservoir—(1) A body of water, either natural or artificial, used for the storage, regulation, and control of water resources. (2) One of the component parts in a **biogeochemical cycle**, where a substance resides; a storage place.

residence time—The interval of time during which a material, such as water substance, remains in a **reservoir**. Often times estimated as the time required to replenish or deplete an amount of substance equal to that presently in the reservoir, assuming constant **inflow** and **outflow** rates.

resistance thermometer—An **electrical-type thermometer** or thermal sensor in which the thermal element is a substance whose electrical resistance varies with the temperature. Such thermometers can be made with very short time constants and are capable of highly accurate measurements. One type, the **thermistor**, is commonly used in **radiosondes**.

resolution—A measure of the ability to separate and distinguish observable quantities.

response time—In meteorology, refers to the ability of an instrument to resolve actual fluctuations in the environmental conditions being monitored. See **time constant**.

resultant wind—A **wind vector** resulting from the addition of several wind vectors.

retention—That portion of the **precipitation** falling on a **drainage area** that does not escape as surface **streamflow** during a specific time interval.

retrogression—In meteorology, the motion of an atmospheric wave or pressure system in a direction opposite to that of the basic flow in which it is embedded.

return period—Same as **recurrence interval**.

return streamer (also known as *main stroke* or *return stroke*)—The intensely luminous discharge that propagates upward from earth's surface to cloud base in the final phase of each **lightning stroke** of a **cloud-to-ground discharge**.

return stroke—Same as **return streamer**.

reversible process—In thermodynamics, changes in one or more variables of a system such that the original state of the system can be restored. Contrast with **irreversible process**.

reversing thermometer—A recording thermal sensor instrument used by oceanographers to measure water temperatures at selected depths; consists of a mercury-in-glass thermometer that records the water temperature when inverted at depth, but retains its reading until the thermometer is returned to its initial position after being brought back to the surface.

ribbon lightning (also known as *band lightning*)—Lightning that appears to spread horizontally into a ribbon when crosswinds slightly displace successive **leaders** of a **lightning flash**.

ridge—Region of the atmosphere in which the atmospheric pressure is high relative to the surrounding region at the same level; usually an elongated area of **anticyclonic** curvature in the wind field. Opposite of **trough** (2).

ridge aloft—Same as **upper-level ridge**.

rime—A white or milky and opaque granular deposit of ice formed by the rapid freezing of supercooled water or fog droplets as they impinge upon a cold exposed object. Rime is denser and harder than **hoarfrost**, but lighter, softer, and less transparent than **glaze**. Rime is composed essentially of discrete ice granules with some crystalline structure, and has densities as low as 0.2 to 0.3 gram per cubic centimeter; glaze is generally continuous but with some air pockets, and has much higher densities.

rip current—A strong surface-water current of short duration flowing seaward from the shore; the return movement of water piled up on the shore by incoming waves and wind. It usually appears as a visible band of agitated water; and, with the outward movement concentrated in a limited band, its velocity is somewhat accentuated.

ripple—Same as **capillary wave**.

rips—The agitation of water caused by interaction of currents and wind; also may be caused by currents moving swiftly over an irregular bottom.

rising limb—The portion of a **hydrograph** indicating an increase in **discharge** resulting from **runoff** of rainfall or snowmelt. This term is most commonly used in discussion of the mathematical or theoretical characteristics of the hydrograph rather than in reference to a specific storm event. Contrast with **falling limb**.

rising stage—The water level (**stage**) in a stream that increases during a time interval.

rising tide (sometimes called ***flood tide***)—The portion of the **tide cycle** between **low water** and the following **high water**. Contrast with **falling tide**.

river basin (also called ***drainage area,*** *watershed*)—The total area drained by a river plus its tributaries.

river forecast—A prediction (**forecast**) of the expected **stage** or **discharge** at a specified time, or of the total volume of flow within a specified time interval, at one or more points along a stream; these forecasts are based upon meteorological and hydrological data.

river gauge—A device for measuring the river **stage**, for example, a **staff gauge**.

river system—The aggregate of stream channels draining a river basin.

roaring forties—A popular nautical term for the stormy ocean regions between 40° and 50° latitude. It nearly always refers to the Southern Hemisphere, where an almost completely uninterrupted belt of ocean in that latitude belt contributes to exceptionally strong prevailing westerly winds.

rocketsonde—An instrument package carried aloft by a **meteorological rocket** to make a **sounding** of the upper atmosphere, at altitudes usually higher than can be obtained by balloon or aircraft.

roll cloud—A low, cyclindrically shaped and elongated cloud occurring behind a **gust front**; associated with but detached from a **cumulonimbus** cloud. Contrast with a **shelf cloud** that is attached to the cloud base. The roll cloud may appear to roll about a horizontal axis.

roll vortex—Circulation within the **planetary boundary layer** about a horizontal axis aligned approximately along the mean wind. The air spirals in opposite directions in neighboring cells. It is often associated with **cloud streets**.

rollers—**Swells** coming across the ocean from a great distance and forming large **breakers** on exposed coasts.

root zone—The layer of soil containing plant roots.

Rossby wave—A series of long wavelength **troughs** and **ridges** in the planetary westerlies; also called **planetary** or **long waves**. Named for Carl-Gustav Rossby (1898–1957), Swedish American meteorologist who studied these waves.

rotary current—A **tidal current** that changes direction progressively through 360° during a **tidal cycle**; typically, rotation is clockwise in the Northern Hemisphere; occurs in the open ocean where the current is not constricted. See also discussion in **current ellipse**.

rotation anemometer—A type of **anemometer** in which the rotation of a sensor element serves to measure the **wind speed**. Rotation anemometers are divided into two classes: those in which the axis of rotation is horizontal, such as a **propeller anemometer**, and those in which the axis is vertical, such as the **cup anemometer**.

rotor cloud—A turbulent, **altocumulus**-type cloud formation found in the lee of some large mountain barriers, particularly in the Sierra Nevada near Bishop, California (see **Bishop wave**). The air in the cloud rotates around an axis parallel to the range; indicative of possible severe **turbulence**. Often improperly called a **roll cloud**.

route forecast—An **aviation weather forecast** for one or more specified air routes. Contrast with **terminal forecast**.

runoff—The water, derived from **precipitation**, that ultimately reaches stream **channels** as **streamflow.**

runoff ratio—The (dimensionless) ratio between the depth of **runoff** to the depth of **precipitation** on a continental surface.

runway visibility—The **visibility** from a particular location along an identified airfield runway; determined by a **transmissometer**. Compare with **runway visual range**.

runway visual range—The maximum horizontal distance down an identified instrument runway of an airfield that a pilot can see and identify standard high intensity lights; determined by a **transmissometer**. This distance is included in **aviation weather observations**, since it is one of the criteria in determining **operational weather limits**.

S

saddle point—Same as **col**.

Saffir–Simpson scale—A hurricane intensity scale developed by H. Saffir (a Dade County, Florida engineer) and R. Simpson (former director of the National Hurricane Center) to relate possible damage caused by **hurricanes** to wind speeds or central atmospheric pressures; 1 is minimal, 5 is most intense.

Saint Elmo's fire—Same as **corona discharge**. A more or less continuous, luminous **electrical discharge** of weak or moderate intensity in the atmosphere, emanating from elevated objects at the earth's surface (lightning conductors, wind vanes, masts of ships) or from aircraft in flight (wing tips, propellers, etc.)

salinity—A measure of the concentration of dissolved salts in **seawater**. Salinity is formally defined as the total amount of dissolved solids in seawater by weight when all the carbonate has been converted to oxide, the bromide and iodide to chloride, and all organic matter is completely oxidized. Salinity is stated in **practical salinity units** (psu) [formerly parts per thousand (‰)]. The average salinity of seawater is 35 (i.e., grams of salt per kilogram of seawater). See **absolute salinity** and **practical salinity**.

salinity current—A **density current** in the ocean whose flow is forced by a relatively more saline (and hence, denser) **water mass** in comparison with surrounding water.

salinometer—Any device or instrument for determining **salinity**, especially one based on electrical conductivity methods.

salt haze—A condition of reduced **visibility** (**haze**) created by the presence of finely divided particles of sea salt in the air, usually derived from the evaporation of sea spray.

saltwater—Any water that has dissolved salts; used to distinguish from **freshwater**. Often used synonymously with **seawater**.

saltwater front—The interface between **saltwater** and **freshwater** in an **aquifer** or **estuary**.

saltwater wedge—A wedgelike intrusion of **seawater** along the bottom of an **estuary**, underlying the near-surface **freshwater**, with a marked increase in **salinity** with depth.

sandstorm—A strong turbulent wind carrying **sand** through the air, the diameter of most of the sand grains ranging from 0.08 to 1 millimeters (0.003 to 0.04 inch). In contrast to a **duststorm**, the sand particles are mostly confined to the lowest 3 meters (10 feet), and rarely rise more than 15 meters (50 feet) above the ground. According to United States observational procedure, a *sandstorm* is reported if **blowing sand** reduces the **horizontal visibility** to less than 1 kilometer (5/8 statute mile), but not less than 500 meters (5/16 statute mile); a *severe sandstorm* is reported if the visibility were reduced to less than 500 meters (5/16 statute miles) due to blowing sand.

Santa Ana Wind—A hot, dry, **foehn type desert wind**, generally from the northeast or east, especially in the pass and river valley of Santa Ana, California, where it is further modified as a **mountain-gap wind**. It blows, sometimes with great force, from the deserts to the east of the Sierra Nevada Mountains and may carry a large amount of dust. It is most frequent in winter; when it comes in spring, however, it does great damage to fruit trees. In autumn its desiccating effects can increase the hazard of wild fires in the canyons of Southern California.

Sargasso Sea—The region (actually in the **horse latitudes**) of the North Atlantic Ocean to the east and south of the **Gulf Stream system**. This is a region of **convergence** of surface waters, and is characterized by clear, warm water, a deep blue color, and large quantities of floating Saragassum or ''gulf weed.''

sastrugi—A series of long, frequently sharp, wavelike ridges of hard snow, characteristic of wind-swept polar plains where the wind tends to blow constantly in one direction. Sastrugi are oriented perpendicular to the wind and have a gentle slope to **windward** and a steep slope to **leeward**.

satellite—A free-flying object that orbits the earth, another planet, or the sun. Artificial satellites in orbit of the earth include **geosynchronous**, **sun synchronous**, and **polar-orbiting satellites**.

satellite dish—Bowl-shaped antenna that collects and focuses the signals that a **satellite** beams down to earth.

satellite image—A representation of the earth and its atmosphere as viewed from a **scanning radiometer** carried on a satellite. See also **visible satellite imagery**, **infrared satellite image,** and **water vapor image**.

satellite meteorology—The study of the atmosphere by means of meteorological data obtained from remote sensors onboard orbiting satellite platforms.

saturated air—**Humid air** in a state of **equilibrium** with a plane surface of pure water or ice at the same temperature and pressure—that is, air whose actual **vapor pressure** is the **saturation vapor pressure** at that temperature; its **relative humidity** is 100%.

saturated soil—Soil whose pores are filled with water. See **zone of saturation**.

saturation—The condition in which the **partial pressure** of any **fluid** constituent is equal to its maximum possible partial pressure under the existing environmental conditions, such that any increase in the amount of that constituent will initiate within it a change to a more condensed state. In molecular-kinetic terms, saturation is attained when the rate of return of molecules of a substance from the dissolved liquid or vapor phase to the more condensed parent phase is exactly equal to the rate of escape of molecules from the parent phase; hence, an **equilibrium** condition. In meteorology, the concept of saturation is applied, almost exclusively, to water vapor as a constituent of the atmosphere.

saturation adiabat—On a **thermodynamic diagram**, a line that describes the change in temperature of a **saturated air parcel** as it is lifted in the atmosphere and undergoes expansional cooling; hence, the curve describes the **saturation-adiabatic lapse rate**.

saturation mixing ratio—The value of the **mixing ratio** of **saturated air** at the given temperature and pressure. This value may be read directly from that mixing ratio line passing through the given temperature and pressure on a **thermodynamic diagram**.

saturation specific humidity—The value of the **specific humidity** of **saturated air** at the given temperature and pressure.

saturation vapor pressure—The partial **vapor pressure** of a system, at a given temperature, wherein the vapor of a substance is in equilibrium with a plane surface of that substance's pure liquid or solid phase; that is, the vapor pressure of a system that has attained **saturation** but *not* **supersaturation**. The saturation vapor pressure of any pure substance, with respect to a specified parent phase, is an intrinsic property of that substance, and is a function of temperature alone.

saturation-adiabatic lapse rate—The rate at which the temperature of a **saturated air parcel** cools with height as it is lifted in a **saturation adiabatic process**; hence, a **process lapse rate** representing the **adiabatic lapse rate** of saturated air. On account of the release of **latent heat** of condensation, its magnitude is less than that of the **dry-adiabatic lapse rate** for unsaturated air and generally between 4° and 7°C per kilometer (2°–3.5°F per 1000 feet). Sometimes called the *moist adiabatic lapse rate*.

saturation-adiabatic process—An **adiabatic process** in which the air is maintained at **saturation** by the **evaporation** or **condensation** of water substance; the **latent heat** associated with the phase change is supplied by or to the air, respectively. The ascent of cloudy air, for example, is often assumed to be such a process. Contrast with the **pseudoadiabatic expansion** process.

savanna—A tropical or subtropical region of grassland and other drought-resistant (xerophilous) vegetation. This type of growth occurs in regions that have a long **dry season** (usually "winter-dry") but a heavy **rainy season**, and continuously high temperatures.

Sc—Abbreviation for the cloud type **stratocumulus**.

scalar—A quantity that can be specified completely with a magnitude, for example, temperature. Contrast with a **vector** quantity.

scan line—A line composed of consecutive **pixels** that is recorded by a **scanning** sensor used by a satellite or radar unit during a single sweep across the field of view.

scanning—In **radar meteorology**, the motion of the radar antenna assembly when searching for **targets**. In satellite meteorology, the sequential motion by a **scanning radiometer** to produce an image.

scanning radiometer—An imaging system aboard a satellite consisting of lenses, moving mirrors, and solid-state image sensors used to obtain radiation fluxes emitted from the earth and its atmosphere; the imaging systems on current operational weather satellites.

scatter—Same as **scattering**; or, sometimes used in referring to the scattered or **diffuse radiation**.

scattered—An official **sky cover** classification for **aviation weather observation** descriptive of a **sky cover** of 0.1–0.5 (5%–54%), applied only when clouds **or obscuring phenomena** aloft are present. Compare with **clear**, **broken**, and **overcast** conditions.

scattering—The process by which small particles suspended in a medium of a different **index of refraction** diffuse a portion of the incident **radiation** in all directions. Contrast with the more organized **reflection**. In scattering, no energy transformation results, only a change in the direction of the radiation. Along with **absorption**, scattering is a major cause of the **attenuation** of radiation by the atmosphere. **Rayleigh** and **Mie scattering** are important types of atmospheric scattering.

scintillation—Generic term for rapid variations in apparent position, **brightness**, or color of a distant luminous object viewed through the atmosphere, caused by changing **refraction** of light passing along an atmospheric path with rapidly changing density. Distinguish between **atmospheric** and **terrestrial scintillation**.

scirocco—See **sirocco**.

scud—Ragged low clouds; most often applied when such clouds are moving rapidly beneath a layer of **nimbostratus**. These clouds are associated with the **cloud species** *stratus fractus* and *cumulus fractus*.

sea—(1) A subdivision of an **ocean**. (2) A term used to describe the portion of the ocean where winds are generating waves within their **fetch** as opposed to **swell**.

sea breeze—A **local wind** regime found along a coast that blows from sea to land, caused by the temperature difference when the sea surface is colder than the adjacent land. Therefore, it usually blows onshore on relatively calm, sunny, summer days and penetrates several kilometers inland by midafternoon; a **sea breeze front** may be apparent. This example of an *onshore breeze* alternates with the oppositely directed, usually weaker, **nighttime land breeze**. As the sea breeze regime progresses, the wind develops a component parallel to the coast, owing to the **Coriolis effect**, causing a **veering** of the wind to the right of the original flow by late afternoon. See **lake breeze**.

sea breeze front—The leading edge of the advancing cool shallow **sea breeze** as it penetrates inland. This **local-scale** feature, which is essentially a **cold front**, can enhance convective clouds and precipitation in the warm air ahead of the front.

sea clutter—A type of **radar echo** returned from the surface, or perhaps just above the surface, of the sea. It may result from any one or a combination of existing conditions such as **spray**, ripples, waves, or **swells**.

sea fog—A type of **advection fog** formed when warm, moist air that has been lying over a warm water surface is transported over a colder water surface, resulting in cooling of the lower layer of air below its original **dewpoint**.

sea ice—Specifically, ice formed by the freezing of **seawater**; often considered as any ice floating in the sea. Contrast with **land ice** of continental origin.

sea level—The datum against which land **elevation** and sea depth are measured. **Mean sea level** is the average of **high water** and **low water** levels, that is, **high** and **low tides**.

sea level pressure—The **atmospheric pressure** at **mean sea level**, either directly measured or, most commonly, empirically determined from the observed **station pressure**. See **reduction to sea level**.

sea salt nucleus—A **condensation nucleus** of a highly **hygroscopic** nature produced by partial or complete desiccation of particles of sea spray or of seawater droplets derived from breaking bubbles on the sea surface.

sea smoke (also known as ***arctic sea smoke***)—Same as **steam fog**, except that it is restricted to over **saltwater** bodies.

sea state—See **state of the sea**.

sea surface temperature (**SST**)—Temperature of the surface layer of a body of water.

season—A division of the year according to some regularly recurrent phenomena, usually astronomical or climatic. See discussions in **winter**, **spring**, **summer**, and **autumn**.

seawater—The water of the seas, distinguished from **freshwater** by its appreciable **salinity**.

Secchi disk—A black and white disk, 30 centimeters or more in diameter, that is lowered into the sea (or lake) on a calibrated line to estimate **transparency** of the water. The depths are noted at which it first disappears when lowered and reappears when raised.

second (s)—The **fundamental unit** of time in **SI** units. For practical purposes, 1/86 400th of a **mean solar day**.

second law of thermodynamics—An inequality asserting that **heat transfer** is not possible from a colder to a warmer system without the occurrence of other simultaneous changes in the two systems or in the environment; simply stated, all systems tend toward disorder or higher **entropy**.

second-trip echo—A **radar echo** received from a **target** beyond the normal maximum range of detection; often associated with **anomalous propagation**.

secondary air pollutant—A substance that is an atmospheric **contaminant** produced by chemical reactions taking place within the atmosphere; **photochemical smog** is an example. Contrast with **primary air pollutant**.

secondary circulation—Atmospheric circulation features of **synoptic scale**. Use of the term is usually reserved for distinguishing between the various dimensions of atmospheric circulation—that is, **primary circulation**, **tertiary circulation**.

secondary cold front—A frontal zone with **cold front** characteristics that forms behind a **frontal cyclone** and the primary cold front within a cold air mass characterized by an appreciable horizontal **temperature gradient**.

secondary cyclone—An atmospheric low pressure system that forms near, or in association with, a primary or older **cyclone**. For example, secondary cyclones often form along the east coast of the United States when a primary cyclone is present in the Great Lakes region. Similarly, secondary cyclones often occur over the Baltic when a primary cyclone is present near the coast of Norway.

secondary front—A **front** that may form within a **baroclinic** cold **air mass,** which itself is separated from a warm air mass by a primary frontal system. The original cold air mass developed different thermal properties in the horizontal because of differences in age or track away from the common **source region**; the resulting contrast may be distinguished as a secondary front.

secondary low—Same as **secondary cyclone**.

secondary rainbow—A **rainbow** of angular radius of approximately 50° about the observer's **antisolar point**; often seen outside the **primary rainbow** of 42° radius, having the color sequence reversed from the primary rainbow. The secondary rainbow results from two internal **reflections** within raindrops plus two **refractions**, one on entering and one on leaving the drop; hence, the secondary bow is less brilliant than the primary bow.

sediment—Particles of organic or inorganic origin that are transported from place of origin and deposited by wind, water, or ice; typically in loose, unconsolidated form.

seiche—Oscillation (having a period from a few minutes to several hours) of the surface of a lake or other small body of water caused by minor earthquakes (seismic seiche), winds, or variations in atmospheric pressure. Usually restricted to a **standing wave** that continues after forcing has ceased.

seismic sea wave—See **tsunami**.

seismograph—An instrument used to measure and record earthquake vibrations and other earth tremors.

selective absorber—A substance that absorbs appreciable radiation only at specific wavelengths or wavelength bands; water vapor and carbon dioxide are examples.

semidiurnal tide—(1) In the oceans, a tide having two **high waters** and two **low waters** each **lunar day**, with little or no **diurnal inequality**. (2) In the atmosphere, a periodic wave with a 12-hour period that describes the largest portion of the diurnal variation of **atmospheric pressure**. See discussion in **lunar atmospheric tide**.

semipermanent pressure systems—Persistent **cyclones** and **anticyclones** that are components of the **planetary-scale circulation**. They are *semipermanent* in that they exhibit some seasonal variations in location and mean surface pressure.

sensible heat—The **heat** absorbed or transmitted by a substance during a change of temperature that is not accompanied by a **change of state**; same as **enthalpy**. Results of sensible heat can be measured with a **thermometer**. Contrast with **latent heat**.

sensible heating—The transport of **sensible heat** (**enthalpy**) from one location or object to another through **conduction**, **convection**, or both, which affects a temperature change.

sensible temperature—The temperature at which ''average indoor air'' of moderate humidity would induce, in a lightly clothed person, the same sensation of comfort as that induced by the actual environment. Compare with **apparent temperature** and **temperature humidity index (THI)**.

sensitivity—In radio terminology, the degree to which a receiver will respond to an input of given strength. The greater the sensitivity, the greater the ability to detect weaker signals.

set (or *current direction*)—The direction toward which an ocean current flows; specified as points of a compass or degrees true north. See **drift** (2).

severe thunderstorm—An intense **thunderstorm** accompanied by locally damaging surface winds (including **tornadoes**), heavy rain, frequent lightning, or large **hail**. The current United States criteria for identifying a severe thunderstorm require that the thunderstorm produces hail 3/4 inch (1.9 centimeters) in diameter or larger and/or winds of 58 miles per hour or greater.

severe thunderstorm warning—Issued by the National Weather Service to warn the public that a thunderstorm producing 3/4 inch (1.9 centimeters) or larger diameter **hail** and/or winds equal to 58 miles per hour or greater is occurring. The warning indicates where the **severe thunderstorm** has been located, what communities are to be affected, and the primary threats to public safety that are associated with the storm. Persons in the affected area should take immediate action.

severe thunderstorm watch—Issued by the National Weather Service to alert the public that conditions are favorable for the development of **severe thunderstorms** in or close to the watch area. Watches typically are in effect for several hours.

severe weather—Any atmospheric condition potentially destructive or hazardous for humans. It is often associated with extreme convective weather (**tropical cyclones**, **tornadoes**, **severe thunderstorms**, **squalls**, etc.) and with storms of **freezing precipitation** or **blizzard** conditions.

sferics—A contraction for **atmospherics**.

shadow—In radar terminology, an **azimuth** sector with no **echo** return because the transmitted signal is blocked by a local prominence (e.g., water tower, building).

shallow-water wave—A wave on the surface of a water body whose wavelength is at least twice the water depth. The speed of this wave type is independent of wavelength, but dependent upon the square root of the water depth; also known as a **long wave** (2). Contrast with a **deep-water wave**.

shear—(1) The rapid variation of a vector **field** over a short distance in space; may be associated with **deformation**. (2) Often used to describe **wind shear**.

shear line—In meteorology, a line or narrow zone across which an abrupt change in the horizontal wind component parallel to this line can be found; a line of maximum horizontal **wind shear**.

sheet frost—A thick coating of **rime** formed on windows and other surfaces.

sheet ice—A smooth thin layer of ice formed on a water surface usually by the coagulation of **frazil**.

sheet lightning—A **lightning flash** (usually a **cloud** or **cloud-to-cloud discharge**) that illuminates clouds so that the clouds have a diffuse but bright white appearance; the lightning flash itself is not visible because it is blocked by clouds.

shelf cloud—A low, wedge-shaped, elongated **accessory cloud** that occurs along a **gust front** often masking the boundary between **updrafts** and **downdrafts**; associated with and attached to the base of a **cumulonimbus cloud**, allowing it to be distinguished from a **roll cloud,** which is not attached. Same as an **arcus cloud**.

shelf ice—Same as **ice shelf**.

shell ice—Ice, on a body of water, that remains as an unbroken surface when the water level drops so that a cavity is formed between the water surface and the ice.

shelterbelt—A belt of trees and/or shrubs arranged as a protection against strong winds; a type of **windbreak**. The trees may be specially planted or left standing when the original forest was cut.

shimmer—Same as **terrestrial scintillation**.

ship report—The encoded and transmitted report of a **marine weather observation**.

shoal—A shallow, submerged feature in a body of water, such as a ridge, bank, or **bar**, that is covered by sediments and poses a hazard to navigation because of its proximity to the surface.

shooting star—Same as **meteor** (2).

shore ice—**Sea ice** that has been beached by wind, tides, currents, or by ice pressure; generally **fast ice**.

shore lead—An open passage between **pack ice** and **fast ice**, or between floating ice and the shore. It may be closed by wind or currents to only a crack.

shoreline—The boundary line between a water body and land, usually taken at mean **high tide**.

short wave—(1) Relatively small, short-wavelength ripples (**troughs** and **ridges**) superimposed on long waves in the **planetary-scale** westerlies; these disturbances propagate with the airflow through the long waves in the middle and upper **troposphere**. (2) In oceanography, same as **deep-water wave**.

short-crested wave—An ocean wave whose **crest** is of finite length, typically on the same order of magnitude as the **wavelength**.

short-range forecast—A weather forecast whose period of validity is less than about three days, lying in the time domain between that covered by **nowcasting** and the **medium-range forecast**.

short-term forecast—A forecast describing hydrometeorological conditions for up to six hours, but typically less (1–2 hours) during active weather.

shortwave radiation—In meteorology, a term used loosely to distinguish radiation in the **visible** and near-visible portions of the electromagnetic spectrum (roughly 0.3–4.0 micrometers in wavelength) from *longwave radiation* (**infrared radiation**).

Showalter stability index—A measure of the local stability of the atmosphere, expressed as a numerical **stability index**. Values of 3 or greater represent stable conditions while values of less than –6 represent **unstable** conditions.

shower—Precipitation from a **convective cloud**. Showers are characterized by the suddenness with which they start and stop, by rapid changes of intensity, and usually by rapid changes in the appearance of the sky.

SI units—The abbreviation for **International System of Units** from the French translation of *Le Système International d' Unités.*

Siberian high—A **cold anticyclone** that forms over Siberia in winter and is particularly apparent on charts of mean sea level pressure. It is centered near Lake Baikal, where the average sea level pressure exceeds 1030 hectopascals (millibars) from late November to early March.

sidereal day—Represents the period of rotation of the earth with respect to a fixed reference frame, such as the stars. By definition, the interval of time between two successive **meridional transits** of the **vernal equinox**; it is equal to 23 hours 56 minutes 4.09054 seconds, which is slightly shorter than the **mean solar day**.

significant level—In a **radiosonde observation**, a level (other than a **mandatory level**) for which values of pressure, temperature, and humidity are reported because temperature and/or moisture-content data at that level are sufficiently important or unusual to warrant the attention of the forecaster, or they are required for the reasonably accurate portrayal of the radiosonde observation.

sill—(1) A submarine ridge that partially separates bodies of water such as seas from one another or from the open ocean restricting the flow of water. (2) An artificial underwater structure built for control of a river.

sill depth—The maximum depth at which horizontal transport of water occurs between an ocean basin and the open ocean.

single-station analysis—The **analysis** (or reconstruction) of the weather pattern from more or less continuous meteorological observations made at a single geographic location; or, the body of techniques employed in such an analysis.

singularity—A weather event that occurs on or near a certain calendar date with unusual regularity; an example is the **January thaw** in New England or the arrival of the monsoon in Arizona in July.

sink—A point, line, or area at which mass or energy is removed from a system, either instantaneously or continuously. Contrast with **source**.

sinking—(1) In atmospheric optics, a **refraction** phenomenon, the opposite of **looming**, in which an object on or slightly above the geographic horizon apparently sinks below it. (2) In oceanography, a **water mass** moving downward. Same as **downwelling** (2).

sirocco (or *scirocco*)—A hot dry south or southeast wind originating from the Sahara and blowing north across North Africa and the Mediterranean Sea, where humidity is increased and then experienced in southern Italy, Sicily, and Malta.

skew *T*-log*p* diagram—The **thermodynamic diagram** with oblique **Cartesian coordinates**: temperature plotted against the logarithm of pressure.

sky—(1) The vaultlike surface above the observer containing airborne objects such as clouds. (2) Same as **state of the sky** or **sky cover**.

sky cover—In **surface weather observations**, a term used to denote one or more of the following: (*a*) the amount of sky covered but not necessarily concealed by clouds or by **obscuring phenomena** aloft; (*b*) the amount of sky concealed by obscuring phenomena that reach the ground; or (*c*) the amount of sky covered or concealed by a combination of (*a*) and (*b*). **Sky cover** is reported in tenths, so that 0.0 indicates a **clear** sky and 1.0 (or 10/10) indicates an **overcast** or a completely covered sky.

sky radiation—Same as **diffuse sky radiation**.

slack water—The condition when a **tidal current** has zero speed; typically occurs during reversal between **ebb** and **flood currents**.

sleet (also known as ***ice pellets***)—In the United States, frozen raindrops that bounce on impact with the ground or other objects. Elsewhere, may refer to a mix of rain and snow, a mix of rain and hail, or melting snow.

slight chance—In a **probability of precipitation** statement within a public forecast, usually equivalent to a 20% chance, with widely scattered areal coverage.

sling psychrometer—A **psychrometer** in which the **wet**- and **dry-bulb thermometers** are mounted upon a frame connected to a handle at one end by means of a bearing or a length of chain. Thus, the psychrometer may be whirled by hand in order to provide the necessary ventilation. Compare with **aspirated** and **Assmann psychrometers**.

slush—Snow or ice on the ground that has been reduced to a soft watery mixture by rain, above freezing temperatures, and/or chemical treatment.

small-craft advisory—Issued by the National Weather Service to advise mariners of sustained (exceeding two hours) weather and/or sea conditions, either present or forecast, potentially hazardous to small boats (less than 65 feet in length). These conditions generally include winds of 18–33 knots and/or dangerous wave conditions. Small craft advisories may be issued also for hazardous sea conditions or lower wind speeds that may affect small craft operations. Advisories can be issued up to 12 hours prior to the onset of adverse conditions.

small-craft warning—Issued by the National Weather Service as a **warning**, for marine interests, of impending winds up to 28 knots (32 mph); used mostly in coastal or inland waters.

smog—A natural **fog** contaminated by industrial pollutants, literally, a mixture of **smoke** and **fog**. Contrast with **dry fog.** Today, it is the common term applied to problematical, largely urban, air pollution, with or without the "natural" fog; however, some visible manifestation is almost always implied.

smoke—Suspended particulate matter in the air resulting from combustion; detected by characteristic odor. The solar disk appears reddish or distant objects have a bluish tinge. In United States observation practice, smoke is a **lithometeor** reported when present either near the surface or in the free atmosphere.

smoothing—An averaging of data in space or time, designed to compensate for random errors or fluctuations of a scale smaller than that presumed significant to the problem at hand.

smudge pot—A device designed for **freeze** or **frost** prevention in an orchard; consists of oil heaters that generate heat and spur convective circulation (**mixing**) of the air.

snap—A colloquial term for a brief period of extreme (generally cold) weather setting in suddenly, as in a "cold snap."

snow—A type of **frozen precipitation** composed of white or translucent ice crystals, chiefly in complex branch hexagonal form and often agglomerated into **snowflakes**, especially at temperatures warmer than –5°C (23°F).

snow accumulation—The collection of snow after any single snowstorm or series of storms. See **snow depth**.

snow advisory—Issued by the National Weather Service to alert the public to an anticipated snowfall that will cause inconvenient conditions but not a threat to life or property if caution is exercised. The exact criteria for snowfall amounts depend on the locale, if the expected snow were to occur at the beginning of the snow season or after a prolonged interval between snow events. Typically, the anticipated amounts for a snow advisory are less than for a **heavy snow warning**.

snow banner (or *snow plume*)—Snow being blown from a mountain crest. It is sometimes mistaken for volcanic smoke or a **banner cloud**.

snow blindness—Impaired vision or temporary blindness caused by sunlight reflected from snow surfaces. The medical name is *niphablepsia.*

snow blink (also known as *snow sky*)—A bright, white glare on the underside of clouds, produced by the multiple reflection of light from a snow-covered surface.

snow burst—An intense convective **snow squall** often accompanied by **lightning** and **thunder**, in what is known as **thundersnow**. Large accumulations of snow can occur, ranging from 1 to 3 inches per hour.

snow core—A cylindrical sample of snow extracted from the **snow cover** by a **snow sampler**. When the core is melted, the snow core represents the **water content** of the snow cover.

snow course—An established traverse of observation points judged to cover representative terrain and **snow cover** for sampling of snow in a **catchment area**; used in **snow surveys**.

snow cover—(1) The areal extent of snow-covered ground, usually expressed as percent of total area in a given region. (2) In general, a layer of snow on the ground surface. (3) The depth of snow on the ground, usually expressed in inches or centimeters. (4) Same as **snowpack**.

snow crust—A crisp, firm, outer surface upon snow, produced by the melting and refreezing of water or by packing by the wind.

snow crystal—Any of several types of ice crystals found in snow. A snow crystal is a single crystal, in contrast to a **snowflake**, which is usually an aggregate of many single snow crystals.

snow density—The mass of **snowpack** per unit volume where mass is the liquid water content of the collected snow sample and the snow volume includes the natural air spaces within the snow sample. Assuming the mass of liquid water is 1000 kilograms per cubic meter, then operationally, the ratio of the volume of **meltwater** that can be derived from a sample of snow to the original volume of the sample; strictly speaking, this is the **specific gravity** of the snow sample.

snow depth—The actual depth of snow on the ground at any instant. See **snow accumulation**.

snow eater—Any warm, dry wind blowing over a snow surface with a resultant rapid disappearance of the snow cover; usually applied to a **foehn** or **Chinook** wind.

snow fence—An open, slatted board fence usually 1–3 meters (3–10 feet) high, placed about 15 meters (50 feet) on the **windward** side of a railroad track or highway. These fences serve to disrupt the flow of the wind in such a manner that the snow is deposited close to the fence on the **leeward** side, leaving a comparatively clear, protected strip parallel to the fence and slightly farther downwind.

snow flurry (or ***flurry***)—Popular term for a **snow shower**, particularly of very light and brief nature. In National Weather Service forecast terminology, snow flurries represent a light snow with no accumulation or a light dusting expected.

snow garland—A rare and beautiful phenomenon in which snow is festooned from trees, fences, etc., in the form of a rope of snow, 2 meters (6.5 feet) or so, and several centimeters in diameter (1 inch), formed and sustained by surface tension acting in thin films of water bonding individual crystals. Such garlands form only when the surface temperature is close to the melting point, for only then will the requisite films of slightly supercooled water exist.

snow gauge—An apparatus designed to measure the amount of a recent **snowfall** by weighing the snow or by melting it.

snow geyser—Fine, powdery snow blown upward by a **snow tremor**.

snow grains (also known as *granular snow*)—A form of **frozen precipitation** consisting of white, opaque particles of ice that are flat or elongated and have diameters less than 1 millimeter (0.04 inch); the solid equivalent of **drizzle**.

snow line—(1) In general, the outer boundary of a snow-covered area. (2) The actual lower limit of the **snowcap** on high terrain at any given time. (3) The ever-changing equatorward limit of snow cover, particularly in the Northern Hemisphere winter. (4) The boundary between the **accumulation zone** and **ablation zone** of a glacier.

snow pellets (formerly called *soft hail* or ***graupel***)—A type of **frozen precipitation** consisting of soft spherical (or sometimes conical) particles of opaque, white ice having diameters of 2–5 millimeters (0.08–0.2 inch). They often break up when striking a hard surface and are distinguished from **snow grains** in being softer and larger. Typically, these represent a **convective** type precipitation in winter.

snow pillow—Pillowlike device filled with antifreeze solution and fitted with a **manometer** that indicates the **water equivalent** of the **snow cover** that accumulates on it.

snow plume—Same as **snow banner**.

snow roller—A cylindrical mass of snow, rather common in mountainous or hilly regions. It occurs when snow, moist enough to be cohesive, is picked up by wind blowing down a slope and rolled onward and downward until it either becomes too large or the ground levels off too much for the wind to propel it further. Snow rollers vary in size from very small cylinders to some as large as a meter long (3 feet) and two meters (6.5 feet) in circumference.

snow sampler—A hollow tube for collecting a sample of snow in situ. See **snow core**.

snow stake—A graduated rule set out to facilitate the measurement of **snow depth**, especially in regions where large **snowfall** accumulations are expected.

snow survey—A systematic determination of the total amount of snow covering a **catchment area** by sampling the **snow cover** along a **snow course**; this survey entails measuring the depth and water content of snow at representative points.

snow squalls—Brief intense **snowshowers** accompanied by strong, gusty winds. Snow accumulation may be significant. Snow squalls are often found in the **lake-effect** region of the Great Lakes.

snow tremor (or *snowquake*)—A disturbance in a **snowfield**, caused by the simultaneous settling of a large area of thick **snow crust** or surface layer. It occurs when wind action has maintained the top foot or more as closely packed, fine-grained snow, rather impervious to air movement; meanwhile, at lower depths **firnification** has caused the larger crystals (**depth hoar**) to grow at the expense of smaller ones, creating air pockets and a weak structure. The collapse of this structure [a perceptible drop, but rarely as much as 2 centimeters (0.75 inches)] may be accompanied by a loud report; and over a large, level field, adjacent patches may settle as a series of tremors.

snowbelt—An area to the lee (downwind) of a large water body such as one of the Great Lakes that is frequently subject to **lake-effect snowstorms**.

snowboard—A flat board of at least 40 centimeters by 40 centimeters that is used to identify fresh **snowfall**. The lightweight board is laid on the previous snow cover and new snow accumulates on the board; the depth of accumulation is measured at specific times and then the board is moved.

snowbreak—Any barrier designed to shelter an object or area from snow (see **snow fence**, **snow shed**); analogous to a **windbreak**.

snowcap—Snow covering the ridges and peaks of mountains when no snow exists at lower elevations.

snowcreep—An extremely slow, continuous, downhill movement of a mass of snow. Contrast with **snowslides (avalanches)**.

snowdrift—The mass of snow deposited behind obstacles or irregularities of the surface, or collected in heaps by eddies in the wind.

snowfall—(1) The rate at which snow falls. In **surface weather observations**, this is usually expressed as inches of snow depth accumulated during a 6-hour interval. (2) Popularly, same as **snow accumulation** for a recent storm. (3) A **snowstorm** where snow is observed to fall.

snowfield—(1) Generally, an extensive area of snow-covered ground or ice, relatively smooth and uniform in appearance and composition. This term is often used to describe such an area in otherwise coarse, mountainous, or glacial terrain. (2) In glaciology, a region of permanent **snow cover**, more specifically applied to the **accumulation zone** of **glaciers**.

snowflake—An **ice crystal** or, much more commonly, an agglomeration of many crystals that falls from a cloud. Contrast with **snow crystal**.

snowmelt—The liquid water resulting from the melting of snow; it may evaporate, seep into the ground, or become a part of **runoff**.

snowpack—The amount of annual accumulation of snow at higher elevations, especially in the western United States; usually expressed in terms of average water equivalent. The **snowmelt** from the snowpack is often managed and used for various purposes.

snowquake—Same as **snow tremor**.

snowshed—A protective structure erected over railroad tracks or mountain roads to prevent snow accumulation on the roadbed. It is used where plowing is difficult, as in deep cuts, or where **snowslides (avalanches)** are frequent.

snowshowers—Snow falling at various intensities for a brief time, typically **convective** in nature. See **shower**. In National Weather Service forecast terminology, snowshowers represent anticipated snow with some accumulation possible.

snowslide—Same as **avalanche**.

sodar—Sound wave transmitting and receiving equipment that emits sound pulses and receives them after they have been scattered back by the atmosphere; used for **sounding** the lower atmosphere for temperature discontinuities and small-scale scattering regions.

soft hail—Same as **graupel** or **snow pellets**.

soft rime—A white, opaque coating of fine **rime** deposited chiefly on vertical surfaces, especially on points and edges of objects, generally in supercooled fog. Contrast with **hard rime**.

soil creep—The slow downslope movement of surface soil or rock debris, usually imperceptible except to observations of long duration. Indicators of this process are tree tilt, or ragged scars in the ground cover. Creep is caused by the combined actions of **gravity**, ground-water flow, and by freezing and thawing or swelling and shrinking of the soil.

soil flow—Same as **solifluction**.

soil moisture—Moisture contained in that portion of the soil that lies above the **water table**, including the water vapor contained in soil pores. Sometimes it refers strictly to the humidity contained in the **root zone** of the plants.

soil moisture deficit—(1) The amount of moisture required to restore a soil sample to **field capacity**. (2) The difference between the water-holding **capacity** of the soil and the actual **soil moisture**.

soil temperature—The temperature of a representative soil type and ground cover made at one or more levels below the surface by special soil thermometers. Measurements are made for agricultural and research purposes.

solar activity—Disturbances on the surface of the sun, such as **solar flares** and **sunspots**, that cause a variation in the appearance or energy output of the sun; may be noticed with disruption of radio transmissions on earth.

solar altitude—The angle of the sun above the local **horizon**; 90° or less; this angle is 90° minus the **solar zenith angle**. See **altitude angle**.

solar constant—The rate at which **solar radiation** is received outside the earth's atmosphere on a surface oriented perpendicular to the incident radiation, and at the earth's mean distance from the sun (149.6 million kilometers); approximately equal to 1376 watts per square meter (2 calories per square centimeter per minute).

solar cycle—Approximate 11-year oscillation in the number of **sunspots** and **solar flares** that affects other solar indices such as the solar output of ultraviolet radiation and the **solar wind**; reckoned from the reversal in the magnetic polarity in sunspots. See **sunspot cycle**.

solar declination—The angular distance (measured in degrees) between the position of the apparent sun on the **celestial sphere** and the **celestial equator**, measured along a **great circle**; see also description in **declination**. This angle varies periodically over the annual cycle, from –23° 27' on 22 December (**winter solstice**) to +23° 27' on 21 June (**summer solstice**).

solar flare—A major disturbance on the solar **chromosphere** seen as a bright eruption of incandescent gas that emits to space high-energy radiation and high-velocity streams of electrically charged subatomic particles. Flares may appear within minutes and fade within an hour. They cover a wide range of intensity and size, and they tend to occur between **sunspots** or over their *penumbrae* (bright regions surrounding a darker *umbra*). Flares may be related to **magnetic storms**.

solar radiation—The total **electromagnetic radiation** emitted by the sun, with 99.0% of the total found in the wavelength band between 0.2 and 4.0 micrometers.

solar spectrum—The part of the **electromagnetic spectrum** between 0.2 and 4.0 micrometers occupied by the wavelengths of **solar radiation**.

solar tide—A **partial tide**, with a period of 12 hours, caused by the **tide-producing force** of the sun.

solar time—Local time as reckoned according to the apparent position of the sun in the observer's sky. Contrast with **civil time**.

solar wind—Stream of ionized gas, mainly hydrogen, continuously flowing outward at variable intensities from the sun at very high speeds into the solar system. While passing near the earth, it interacts with earth's **magnetic field** and produces various effects in the high atmosphere (e.g., **aurora**).

solar zenith angle—The angle measured at the earth's surface between the sun and the local **zenith** along a **great circle** including that zenith and the sun; this angle is 90° minus the **solar altitude angle**. See **zenith angle**.

solid—One of the three basic physical phases of matter; a substance that does not flow under moderate stress. Compare with **gas** and **liquid**.

solid rotation—The rotation of a system as though it were a **solid** or rigid body rotating about a fixed axis, all points within the body having the same **angular velocity**.

solifluction (or *soil flow*)—The rapid creep and/or viscous downslope flow of saturated soil. The process commonly takes place in high latitudes at high elevations, initiated and augmented by **frost action**.

solstice—(1) Either of the two points on the sun's apparent annual path where it is displaced farthest, north or south, from the earth's **equator**—that is, a point experiencing the greatest deviation (**declination**) of the ecliptic from the **celestial equator**. (2) Popularly, the time at which the sun is farthest north or south; the "time of the solstice." In the Northern Hemisphere the **summer solstice** falls on or about 21 June, and the **winter solstice** on or about 22 December. The reverse is true in the Southern Hemisphere.

solute effect—A reduction in the **relative humidity** required for **condensation** caused by **hygroscopic** substances (e.g., sea salt crystals) dissolved in pure water; this may aid in droplet growth.

Somali Current—An ocean current in the Indian Ocean flowing southwestward along the coast of Somalia (East Africa) during part of the year as a continuation of the **North Equatorial Current**. In summer (Northern Hemisphere), when the **North Equatorial Current** and the **Equatorial Countercurrent** are replaced by an eastward-flowing **Monsoon Current**, the Somali current reverses its direction and flows north from about 10°S.

sonar (*so*und *na*vigation *a*nd *r*anging)—An instrument that utilizes the propagation of underwater **sound waves** to locate and detect the type of underwater objects.

sonic anemometer—An instrument sensor that emits **sound waves** to determine **wind speed**. The speed of propagation of a sound wave consists of the vector sum of the speed of sound in air at the measured temperature and the wind speed.

sonora—A summer thunderstorm in the mountains and deserts of southern California of the United States and Baja California, Mexico.

sound—A periodic variation in the pressure or particle motion in an elastic medium, such as the atmosphere or a water body. See also as **sound wave**. Humans detect sounds that have frequencies within the **audio frequency** range (20–20 000 Hertz).

sound wave—Same as **acoustic wave**.

sounder—A special type of **radiometer** that measures changes in atmospheric temperature with altitude, as well as the content of various chemical species in the atmosphere at various levels. The *High Resolution Infrared Radiation Sounder (HIRS)* flown aboard NOAA polar-orbiting satellites is an example.

sounding—A probe of the environment. (1) In meteorology, same as **upper-air observation** as taken from a **radiosonde**, **rocketsonde**, **dropsonde**, or satellite **sounder**. However, a common connotation is that of a single complete **radiosonde observation** (measurement of atmospheric conditions aloft). (2) The measurement of the water depth beneath a vessel by line or other devices.

sounding balloon—A free, unmanned balloon instrumented and/or observed for the purpose of obtaining a sounding of the atmosphere. See also **radiosonde**.

source—(1) A point, line, or area at which mass or energy is added to a system, either instantaneously or continuously. Contrast with **sink**. (2) Origin of a river.

source region—Extensive region of the earth's surface characterized by essentially uniform surface conditions and so located with respect to the general atmospheric circulation pattern that a volume of air remains in contact with the surface long enough to acquire properties that distinguish it as an **air mass** (e.g., central Canada for continental polar air). See **airmass source region**.

South Atlantic Current—An eastward-flowing current of the South Atlantic Ocean at about 40°S that is continuous with the northern edge of the **Antarctic Circumpolar Current**.

South Atlantic high—A semipermanent anticyclone located in the **subtropical high-pressure belt** of the South Atlantic Ocean.

South Equatorial Current—Any of several ocean currents driven by the southeast **trade winds** flowing over the tropical oceans of the Southern Hemisphere. In the Atlantic Ocean it flows westward between the equator and 20°S. Part crosses the equator and flows northwest along the coast of South America as the **Guiana Current**. The rest turns to the left and flows south along the coast of Brazil as the **Brazil Current**. In the Pacific Ocean, the **South Equatorial Current** crosses the ocean from east to west between the latitudes of approximately 3°N and 10°S. Much of it turns south in midocean forming a large anticyclonic whirl. The portion that continues across the ocean divides as it approaches Australia, part moving north toward New Guinea and part turning south along the east coast of Australia as the **East Australia Current**. In the Indian Ocean, the **South Equatorial Current** is displaced rather far to the south, and as it approaches the east coast of Africa it turns south, joining the **Agulhas Current**.

South Indian Current—An eastward-flowing ocean current in the South Indian Ocean, along the northern edge of the **Antarctic Circumpolar Current**; the South

Indian Current is an extension of the **Agulhas Current** and joins the **West Australian Current** between New Zealand and Chile.

South Pacific Current—An eastward-flowing current of the South Pacific Ocean that is continuous with the northern edge of the **Antarctic Circumpolar Current**.

South Subtropical Current—A warm westward-flowing current in the South Pacific, found to the south of the South Equatorial Current at approximately 15°–20°S latitude; merges with the **East Australian Current**.

southeast trades—The **trade winds** of the Southern Hemisphere, with a **wind direction** from the southeast. This belt is usually located between 0° and 30°S, on the equatorward flank of the **subtropical highs** of the Southern Hemisphere.

southeaster—A southeasterly wind (blowing from the southeast), particularly a strong wind or gale; for example, the winter southeast storms of the Bay of San Francisco.

Southern Oscillation—Opposing swings of surface air pressure between the eastern and western tropical Pacific Ocean; associated with intense **El Niño** events. These pressure variations, monitored in Tahiti and in Darwin, Australia, appear to affect the **Walker Circulation**, the distribution of mass between hemispheres and changes in wind, temperature, and rainfall regimes.

southwest monsoon—(1) The high sun **season** (Northern Hemisphere summer) with prevailing southwest onshore winds over south Asia; typically, the **rainy season** in India. (2) Current of warm **humid air** that affects the southwest region of the United States in July and August; sometimes called the *Arizona monsoon*.

special observation—A category of **aviation weather observations** taken and transmitted as an unscheduled coded message to report significant changes in one or more **weather elements** since the previous **record observation**. Certain criteria requiring a special observation include changes in **ceiling height**, **visibility**, and wind, as well as the occurrence of **thunderstorms**, **tornadoes**, and **freezing** or **frozen precipitation**.

specific energy—In hydrology, the sum of the **elevation** of the free surface above the bed and the velocity **head**, based on the mean velocity at that section; hence, the energy at any cross section of an open channel, measured above the channel bottom as datum.

specific gravity—The (dimensionless) ratio of the mass **density** of substance to the density of pure water usually at 4°C (39.2°F), the temperature of maximum density of liquid water. Specific gravity should not be confused with the *acceleration of gravity*, which is the force of **gravity** per unit mass.

specific heat—The **heat capacity** of a system per unit mass; the amount of heat required to raise the temperature of 1 gram of a substance by 1 Celsius degree. A common unit is calories per gram per Celsius degree.

specific humidity—In a system of **humid air**, the (dimensionless) ratio of the mass of water vapor to the total mass of the system, usually expressed as grams of water vapor per kilogram of humid air.

specific volume—Volume per unit mass of a substance, the reciprocal of mass **density**.

specific yield—The quantity of water that a unit volume of **aquifer**, after being saturated, will yield when drained by **gravity**. It is expressed either as a ratio or as a percentage of the volume of the aquifer. Specific yield is a measure of the water available to wells.

spectral band—A finite segment in the electromagnetic spectrum bounded by a set of wavelengths; for example, radiation in the **visible** spectral band has wavelengths between 0.39 and 0.76 micrometer.

specular reflection—A highly directed **reflection** of radiation from a smooth, "mirror-like" surface, where the **angle of reflection** equals the **angle of incidence**. Contrast with **diffuse reflection**.

speed of light—The speed of propagation of **electromagnetic radiation** of any wavelength through a "perfect vacuum." It is a universal dimensional constant equal to 2.997925 ^{8} meters per second (670 million mph).

speed of sound—The speed with which a **sound wave** advances through a medium. The speed of sound varies widely from one medium to another and depends upon the physical state of any given medium, in particular, the temperature, but this is not strongly dependent upon the frequency of the sound wave. The speed of sound at sea level in the **standard atmosphere** (15°C and 1013 hectopascals) is 340 meters per second (760 mph).

spicules—Small, needlelike crystals of ice (see **ice needles**), often times, newly formed **sea ice**.

spillover—(1) That part of **orographic precipitation** that is carried along by the wind so that it reaches the ground in the nominal rain shadow on the lee side of the barrier. (2) Flow of water from a reservoir over or through a spillway.

spinup vortices—Small **eddies** that form beneath **thunderstorms** in an **eyewall** of a **hurricane (typhoon)**; these vortices may add to the surface winds of the system to produce the strongest winds of a hurricane.

spiral band—The characteristic arrangement of **radar echoes** received from precipitation areas within intense **tropical cyclones** (**hurricanes** or **typhoons**). These bands of individual echoes curve **cyclonically** in toward the center of the storm and appear to merge to form the wall around the **eye** of the storm, forming a spiral shape.

spiral layer—Same as **Ekman layer**.

Spitzbergen-Atlantic Current—A warm branch of the **Norway Current** that flows northward toward Spitzbergen in the Arctic Ocean, then westward toward Greenland to join with the **East Greenland Current**.

splash—Dispersion of **raindrops** reaching the earth's surface and rebounding. Proper placement of **rain gauges** should minimize splash.

split-flow pattern—In the upper-level planetary-scale circulation, westerlies to the north appear to be detached from and have a different wave pattern from westerlies to the south.

Spörer minimum—A period of reduced **sunspot** activity from about 1450 to 1550 A.D.

spray—Ensemble of water droplets torn by the wind from the surface of an extensive body of water, generally from the **wave crests**, and carried up a short distance into the air.

spring—(1) The **season** of the year comprising the transition period from **winter** to **summer**; the *vernal* season, during which the sun is approaching the **summer solstice**. In popular usage and for most meteorological purposes, spring is customarily taken to include the months of March, April, and May in the Northern Hemisphere, and September, October, and November in the Southern Hemisphere. Except in the Tropics, spring is a season of rising temperatures and decreasing cyclonic activity over continents. In much of the Tropics, neither spring nor fall is recognizable, and in polar regions, both are very short-lived. (2) Astronomically, the period extending from the **vernal equinox** to the **summer solstice**. (3) A place where **groundwater** flows naturally from a rock outcrop or soil.

spring equinox—Same as **vernal equinox**; the equinox at which the sun approaches the Northern Hemisphere; the time of occurrence is approximately 21 March.

spring snow—A coarse, granular, wet snow, resembling finely chopped ice, generally found in the spring.

spring tide—A large-amplitude **tide** near the time of **syzygy** or twice monthly (new and full moon phases), when **tidal ranges** between **high water** and **low water** are greatest. Contrast **neap tide**.

sprinkle—Popular term for a very light **shower** of rain with essentially no accumulation.

squall—(1) A strong wind characterized by a sudden onset, a duration of the order of minutes, and a rather sudden decrease in speed. In United States observational practice, a squall is reported only if a **wind speed** of 16 knots (18 mph) or higher were sustained for at least two minutes (thereby distinguishing it from a **gust**). (2) A severe local storm considered as a whole, that is, winds and cloud mass and (if any) precipitation, thunder, and lightning.

squall line—Any nonfrontal line or narrow band of active **thunderstorms** (with or without **squalls**); a mature **instability line**. These lines may be of considerable length.

SST—Abbreviation for **sea surface temperature**.

St—Abbreviation for the cloud type **stratus**.

stability—(1) A property of a system where disturbances induced into the system will decrease in magnitude and the system returns to the original steady state. Contrast with **instability**. (2) The property of the atmosphere, usually assumed in **hydrostatic equilibrium** with respect to vertical displacements of **air parcels**. Using the **parcel method**, stability would indicate a situation where a displaced parcel would return to its origin, or, in general, a disturbance would diminish. The criterion for stability is based on a comparison of the **environmental lapse rate** with either the

dry-adiabatic lapse rate (for unsaturated parcels of air) or the **saturation-adiabatic lapse rate** (for saturated parcels of air). (3) In the oceans, the resistance to vertical **mixing** or overturning of **water masses** because less dense water lies above more dense water—that is, a positive vertical density gradient.

stability index—An indication of the local stability of a layer of air. One or more of these indices are often used in severe weather forecasting. Operational indices include the **Showalter index, K-index,** or **Lifted Index (LI)**.

stable air—An air mass in which **stability** prevails such that vertical displacements are suppressed, a condition that depends on the vertical gradients of air temperature and humidity. Contrast with **unstable air**.

stable motion—Motion in which small disturbances do not grow. Contrast with **unstable motion**.

staff gauge—A graduated scale with water-level markings placed in a position so that the **stage** of a stream may be read directly therefrom; a type of **river gauge**. Staff gauges may be painted on bridge piers and pilings, and painted wooden scales or enameled metal scales may be mounted on bridges, piers, docks, trees, or specially constructed supports to be partially submerged by the stream at all times.

stage—(1) The **elevation** of the water surface in a stream as measured by a **river gauge** with reference to some arbitrarily selected zero datum. (2) In geology, a time interval, especially a primary subdivision of a glacial epoch.

stage-discharge relation—Same as **rating curve**.

stagnation area—In pollution studies, a region of the surface layer where the following conditions persist for at least four days: a **geostrophic wind** less than 15 knots (17 mph), no frontal passage and no precipitation.

standard atmosphere—(1) A hypothetical vertical distribution of atmospheric temperature, pressure, and density, which, by international agreement, is taken to be representative of the atmosphere for purposes of **pressure altimeter** calibrations, aircraft performance calculations, aircraft and missile design, ballistic tables, etc. (2) A standard unit of **atmospheric pressure**, defined as that pressure exerted by a 760-millimeter column of mercury at **standard gravity** (9.80665 meters per second squared) at a temperature of 0°C (32°F). The recommended unit for meteorological use is 1013.25 hectopascals (millibars).

standard depth—A depth below the sea surface at which water properties should be measured and reported, either directly or by interpolation, according to the proposal by the International Association of Physical Oceanography in 1936; analogous to the **mandatory levels** of meteorological **upper-air observations**. Several accepted standard depths, in addition to the surface, include 10, 20, 30, 50, 75, and 100 meters.

standard gravity—A reference value of the sea level acceleration of **gravity;** used in the *gravity correction* of **mercury barometers**. The accepted value in the standard atmosphere is 9.80665 meters per second squared for a latitude of 45.5425°N.

standard pressure—(1) In meteorology, the arbitrarily selected **atmospheric pres-**

sure of 1000 hectopascals (millibars) to which **adiabatic processes** are referred for definitions of **potential temperature**, etc. (2) In physics, a pressure of one **standard atmosphere** (1013.25 hectopascals).

standard rain gauge—A type of **nonrecording rain gauge** approved for official use in a national weather-observing network. In the United States, the most common standard gauge has an 8-inch diameter collector funnel and a collector tube designed to amplify the **catch** by ten-fold for ease of reading.

standard temperature and pressure (STP)—A phrase used in physics to indicate a temperature of 0°C (32°F) and a pressure of one **standard atmosphere** (1013.25 hectopascals).

standing cloud—Any stationary cloud maintaining its position with respect to a mountain peak or ridge, such as a **banner cloud**, **cap cloud**, **crest cloud**, or a **lenticular cloud**.

standing wave—(1) In meteorology, the same as **stationary wave**. (2) In oceanography, a nonprogressive wave type where the water surface oscillates vertically between fixed points, or **nodes**. The nodes are at points where no vertical displacement of the water surface occurs. In between, the maximum vertical displacement of the standing wave occurs at the **antinodes**. (3) See also **hydraulic jump**.

state—(1) The physical phase of matter: solid, liquid, or gas. (2) The specified conditions of a system, such as those thermodynamic conditions described mathematically in the **equation of state** of a gas by the **variables of state** (temperature, pressure, and mass density). See also **change of state.**

state of the sea (or *sea state*)—A standardized description of the properties of the wind-generated waves on the surface of the sea; also used to estimate **wind speed** from the **Beaufort scale**.

state of the sky—The characteristics of the clouds present in the atmosphere at a specific moment (amount, genus, height, etc.).

static pressure—The component of the force measured perpendicular to the flow per unit area oriented parallel to the direction of **fluid** flow. Alternately, the pressure measured by a **barometer** moving with the fluid. Contrast with **dynamic** and **hydrostatic pressure**.

static water level (or *static level*)—In hydrology, the height of the **piezometric surface** or the height to which water will rise in an **artesian well**. The static level of a flowing well is above the ground surface.

station elevation—The officially designated vertical distance above **mean sea level** that is adopted as the reference datum level for all current measurements of **atmospheric pressure** at the station; this elevation is usually the same as the **airport elevation**.

station model—A specified pattern for entering, on a **weather map**, the meteorological symbols that represent the state of the weather at a particular observing station.

station pressure—The **atmospheric pressure** computed for the level of the station elevation; contrast with **sea level corrected pressure**.

stationary front—A zone of transition between contrasting **air masses** that does not move or is almost motionless; winds blow parallel to the **front** but in opposite directions on either side of the front. See **quasi-stationary front**.

stationary wave (also known as ***standing wave***)—A nonprogressive wave such that the points maintaining zero amplitude (**nodes**) will remain stationary relative to the given coordinate system. In meteorology, the coordinate system is usually fixed with respect to the earth, so that a stationary wave usually refers to a wave that is stationary relative to the earth's surface. Often observed in the vicinity of above and downwind of mountain ranges.

steady state—A **fluid** motion in which the velocities at every point of the field are independent of time. Sometimes an assumption is made that all other properties of the **fluid** (pressure, density, etc.) are also independent of time.

steam fog—The general name for a ground-level cloud produced when cold air comes in contact with a relatively warm water surface; has the appearance of rising streamers. See **arctic sea smoke**.

steam point—The true **boiling point** of water, defined as the temperature of an equilibrium mixture of liquid water and steam at one **standard atmosphere pressure** (1013.25 hectopascals), or 100°C (212°F); a fiducial point on the **Celsius** or **Fahrenheit temperature scales**.

steering—In meteorology, loosely used for any influence upon the direction of movement of an atmospheric disturbance exerted by another aspect of the state of the atmosphere.

steering flow—A basic flow that exerts a strong influence upon the direction of movement of disturbances embedded in it; this is the usual concept of **steering** in meteorology.

steering level—A level, in the atmosphere, where the velocity of the basic flow bears a direct relationship to the velocity of movement of an atmospheric disturbance embedded in the flow.

Stefan–Boltzmann law—A fundamental mathematical relationship of electromagnetic radiation that states that the amount of energy radiated per unit time from a unit surface area of an ideal **blackbody** is proportional to the fourth power of the absolute temperature of the blackbody. The law is written: $E = \sigma T^4$, where E is the **emittance** of the blackbody in units of watts per square meter, σ is the *Stefan–Boltzmann constant* (5.6696×10^{-8} watts per square meter per the fourth power of kelvin), and T the absolute (Kelvin) temperature of the blackbody. Named for two Austrian physicists, Josef Stefan (1835–1893) and Ludwig Boltzmann (1844–1906).

steppe—An area of grass-covered and generally treeless plains, with a semiarid climate, which forms a broad belt over southeastern Europe and the southwestern part of the former Soviet Union.

stepped leader—The initial **streamer** or **leader** stroke of a first discharge of **lightning** in which the ionized channel is propagated downward by successive jumps.

Stevenson screen—A type of **instrument shelter**. The shelter is a wooden box painted white with double louvered sides and mounted on a stand four feet above the ground. In addition to the **dry-** and **wet-bulb thermometers**, it usually contains **maximum** and **minimum thermometers**.

storage—The quantity (mass or volume) of a substance that is impounded in a **reservoir** of the **biogeochemical cycle**; usually liquid water in the **hydrological cycle** that is impounded in the soil or a water body.

storm—(1) Any disturbed state of the atmosphere, especially affecting the earth's surface, and strongly implying potentially destructive or otherwise unpleasant weather. (2) An atmospheric disturbance involving perturbations of the prevailing pressure and wind fields, on scales ranging form tornadoes (1 kilometer or 0.5 mile) to extratropical cyclones (2000–3000 kilometers or 1500 miles across). (3) In the modern international **Beaufort wind scale** (*Beaufort force number 10*), a wind with a speed between 48 and 55 knots (55–63 mph), capable of considerable structural damage and making the sea white with foam.

storm model—A physical, three-dimensional representation of the inflow, outflow, and vertical motion of air and water vapor in a storm.

storm surge—An abnormal local rise in sea level accompanying a **tropical cyclone** or other intense storm and whose height is the difference between the observed level of the sea surface and the level that would have occurred in the absence of the storm. Storm surge is usually estimated by subtracting the elevation of the normal or astronomical tide from that of the observed **storm tide**. A storm surge, typically lasting for several hours, is potentially hazardous along low lying coasts.

storm tide—The actual level of seawater resulting from the normal or astronomical tide combined with the **storm surge**.

storm warning—(1) Meteorological message intended to warn those concerned of the occurrence or expected occurrence of a wind of Beaufort force 10 or 11 over a specific area. (2) Any forecast of severe weather conditions. The National Weather Service issues storm warnings to warn mariners of current or predicted 1-minute sustained surface winds of 48 knots or greater; these warnings may be issued up to 24 hours prior to the onset of these conditions. These storm warnings do not cover those conditions directly associated with **tropical cyclones**.

storm wind—In the **Beaufort wind scale**, a wind whose speed is from 56 to 63 knots (64 to 72 mph).

STP—Abbreviation for **standard temperature** and **pressure**.

straight line winds—Typically used to describe thunderstorm winds with little rotational wind damage to differentiate from the winds in a **tornado** that produce damage that exhibit rotational characteristics. These straight line winds are most often found with a **gust front**, originating from **downdrafts**.

stratification—The formation or existence of distinct layers in a **fluid** (the atmosphere or large bodies of water) identified by distinct differences in an observed quantity such as temperature, density, salinity, etc.

stratiform—*Like stratus*; descriptive of clouds of extensive horizontal development, as contrasted to the vertically developed **cumuliform** types; typical of nonconvective lifting processes. **Stratus**, **altostratus**, and **cirrostratus** are examples of stratiform clouds.

stratocumulus (**Sc**)—A principal **low cloud** type (**cloud genus**), predominantly **stratiform**, in the form of a gray and/or layer or patch, which nearly always has dark parts and is nonfibrous (except for **virga**). Its elements are tessellated, rounded, roll-shaped, etc.; they may or may not be merged, and usually are arranged in orderly groups, lines, or undulations, giving the appearance of a simple (or occasionally a cross pattern) wave system. These elements are generally flat-topped, smooth, and large. Stratocumulus is composed of small water droplets, sometimes accompanied by larger droplets, **soft hail**, and (rarely) by snowflakes. When the base of stratocumulus is rendered diffuse by precipitation, the cloud becomes **nimbostratus**. Thin stratocumulus may also have **corona** phenomena.

stratopause—The top of the **inversion layer** in the upper **stratosphere** at an altitude of about 50 kilometers (31 miles);a relative temperature maximum, with temperatures reaching 5°C just below the **mesosphere**.

stratosphere—The thermal subdivision of the atmosphere situated between the **troposphere** and **mesosphere**; primary site of **ozone** formation. Within the stratosphere, air temperature is constant with altitude (isothermal) at lower levels (10–20 kilometers) and then increases with increasing altitude (20–50 kilometers), culminating in a temperature maximum at the **stratopause**.

stratospheric coupling—The interaction between disturbances in the **stratosphere** and those in the **troposphere**.

stratospheric warming—Temporary or permanent breakdown of the **antarctic or arctic stratospheric vortex**, in late winter or early spring, due to a rapid temperature rise of the polar stratosphere [up to about 50 K (90 F°) in a few days].

stratus (**St**)—A principal **low cloud** type (**cloud genus**) in the form of a gray layer with a rather uniform base. Stratus does not usually produce precipitation, but when it does occur it is in the form of minute particles, such as **drizzle**, **ice crystals**, or **snow grains**. Stratus often occurs in the form of ragged patches, or cloud fragments (*stratus fractus*, see **scud**). When the sun is seen through the cloud, its outline is clearly discernible, and it may be accompanied by **corona** phenomena. The particulate composition of stratus is quite uniform, usually of fairly widely dispersed water droplets, and at lower temperatures, of ice crystals (although this is much less common).

streak lightning—Ordinary lightning, of a **cloud-to-ground discharge**, that appears to be entirely concentrated in a single, relatively straight lightning channel.

stream gauge—Same as **river gauge**.

streamer—The initial high ion density channel in a **lightning discharge** that establishes a conductive path through the atmosphere. A **stepped leader**, **dart leader**, and **return streamer** are special types of streamers.

streamflow—The water flowing in a stream or river channel.

streamline—The pattern of flow of air moving horizontally; on a map, lines that are drawn everywhere parallel to **wind direction**; also used to represent the flow of **groundwater**.

stretching—A change in the shape of a **fluid** through **deformation** so that the **streamlines** of fluid flow diverge along one directional axis, while converging along a perpendicular axis. Stretching is an important feature in leeside **cyclogenesis** and **tornado** formation.

stress—An external force per unit area acting on a surface as a consequence of the **transport** of **momentum** by turbulent flow. See **wind stress**.

strong breeze—In the **Beaufort wind scale** (*Beaufort force number 6*), a wind whose speed is from 22 to 27 knots (25 to 31 mph), causing movement of large tree branches and making large waves with some spray.

strong gale—In the **Beaufort wind scale** (*Beaufort force number 9*), a wind whose speed is from 41 to 47 knots (47 to 54 mph), producing slight structural damage and causing a roll over of wave crests.

Stüve diagram—A **thermodynamic diagram** with temperature as abscissa and pressure to the power 0.286 as ordinate, increasing downward. The (**dry**) **adiabats** are thus straight lines radiating from the point *T (temperature)* = 0 K, *P (pressure)* = 0 (not diagrammed). **Pseudoadiabats** and **saturation mixing ratio** lines have a pronounced curvature.

subarctic climate—Same as **taiga**.

Subarctic Current—Same as **Aleutian Current**.

subgeostrophic—Any **wind speed** less than the hypothetical **geostrophic wind** speed calculated for that **latitude** and horizontal **pressure gradient**.

subjective analysis—A technique for the interpretation of the two-dimensional distribution of field variables involving the drawing of isopleths based upon human experience and checking of spatial and or temporal **continuity**; this **analysis** technique does not employ the sophisticated numerical interpolation techniques utilized by **objective analysis**.

sublimation—The transition of a substance from the solid phase directly to the vapor phase, without passing through an intermediate liquid phase. Opposite of **deposition**. **Latent heat** of sublimation is extracted from the system and stored in the vapor.

subpolar anticyclone—Same as **subpolar high**.

subpolar glacier—A polar **glacier** with 10–20 meters (33–65 feet) of **firn** in the **accumulation zone** where some melting occurs.

subpolar high—An **anticyclone** that forms over the cold continental surfaces between the latitudes of 50° and 70°, principally in Northern Hemisphere winter; these high pressure cells typically migrate eastward and southward.

subpolar low-pressure belt—A belt of low **atmospheric pressure** located, in the mean, between 50° and 70° latitude. In the Northern Hemisphere, this ''belt'' consists of the **Aleutian low**, and the **Icelandic low**. In the Southern Hemisphere, several low pressure cells are found to exist around the periphery of the Antarctic continent in the Southern Oceans.

subsatellite point—Location where a straight line drawn from a **satellite** to the center of the earth intersects the earth's surface. See discussion in **ground track**.

subsidence—Slow descent of a mass of air, over a wide area, generally accompanied by horizontal divergence in the lower layers. The subsiding air is compressed and warmed by **adiabatic processes** and its initial stability is generally increased.

subsidence inversion—A **temperature inversion** produced by the **adiabatic warming** of a layer of sinking (subsiding) air. This inversion may be enhanced by vertical **mixing** in the air layer below the inversion. The **trade wind inversion** is a subsidence inversion.

subsun—A **photometeor** consisting of a bright elongated spot of light appearing below the sun. Typically, the observer must be elevated above an aggregate of ice crystals, as in an airplane. The subsun represents a downward extension of a **sun pillar**, caused by **reflection** of sunlight off ice crystals suspended in the air.

subsurface current—A movement of water usually flowing below the **thermocline**, at speeds slower than the **surface currents** above; the current directions may also differ. The **Equatorial** or **Cromwell Undercurrent** is an example.

subsurface flow—Any flow of water below the earth's surface, to include **interflow** plus **groundwater** flow.

subtropical anticyclone—Same as **subtropical high**.

subtropical climate—In general, a **climatic zone** with a climate typical of the **subtropics**, with warm temperatures and with meager precipitation.

Subtropical Convergence—An oceanic zone separating ocean currents coming together (converging) and carrying subtropical and cold subpolar water masses. Typically refers to the well-defined zone in the southern oceans at about 40°S, but also to the more poorly defined zone in northern oceans between 20° and 40°N.

subtropical easterlies—Same as **tropical easterlies**.

subtropical high—One of the semipermanent **anticyclones** of the **subtropical high-pressure belt**. They appear as **centers of action** on charts of mean surface pressure. They lie over oceans, and are best developed and displaced poleward in the high sun (summer) season. Examples include the **Azores high**, **Bermuda high**, **Pacific high**, and **South Atlantic high**.

subtropical high-pressure belt—One of the two belts of high atmospheric pressure that are centered, on average, near 30°N and 30°S. These belts are formed by the **subtropical highs**.

subtropical jet stream—A zone of unusually strong upper-level winds situated between the tropical **tropopause** and the midlatitude tropopause.

subtropical ridge—Same as **subtropical high-pressure belt**; or the axis thereof.

subtropics—The indefinite belts in each hemisphere between the **tropic** and **temperate zones**. The poleward boundaries are considered to be roughly 35° N and S, but vary greatly according to continental influence, being farther poleward on the west coasts of continents and farther equatorward on the east coasts.

suction vortices—A group of **whirlwinds** (typically 3–5) that sometimes accompany a **tornado**; they hang down from the base of the tornado-producing cloud and revolve around and in the same direction as the winds in the main funnel. These vortices may produce cyclodial ground markings seen from aloft.

sudden ionospheric disturbance (SID)—An abrupt and temporary (15–30 minutes) increase in the rate of **ionization** in the **ionosphere** triggered by a surge in **solar activity** (**solar flare**); interrupts radio communication.

summer—(1) The warmest **season** of the year everywhere except in some tropical regions; the "high sun" season during which the sun is most nearly overhead (i.e., at **solstice**). Popularly and for most meteorological purposes, summer is taken to include June, July, and August in the Northern Hemisphere, and December, January, and February in the Southern Hemisphere; the reverse of **winter**. (2) Astronomically, the period extending from the Northern Hemisphere **summer solstice**, about 21 June, to the **autumnal equinox**, about 22 September.

summer monsoon—A seasonal wind regime (**monsoon**) of oceanic origin that blows onshore during the high sun (summer) season. Opposite of **winter monsoon**. The **southwest monsoon** in South Asia is an example of the summer monsoon.

summer solstice—The point on the **celestial sphere** where the sun reaches its greatest distance (**declination**) north or south from the **celestial equator** for the Northern or Southern Hemispheres, respectively. Opposite the **winter solstice**. In the Northern Hemisphere, this event occurs approximately on 21 June, when the sun appears overhead on the **Tropic of Cancer**.

sun dog—Same as **parhelion**.

sun glint—A **specular reflection** of direct rays of the sun by a water surface; seen as an illuminated region of the dark ocean surface on **visible satellite imagery**, with the size and brightness depending upon the **state of the sea**.

sun pillar—A **photometeor** consisting of a luminous streak of light, white or slightly reddened, extending above and below the sun, most frequently observed near **sunrise** or **sunset**; it may extend to about 20 degrees above the sun, and generally ends in a point. Acknowledged as caused by **reflection** of sunlight off ice crystals suspended in the air.

sun synchronous satellite—An artificial **satellite** placed in an orbit about the earth with an orbital plane fixed relative to the sun. This type of satellite passes over each point on the earth's surface under its path at the same times daily, once on the ascending node and once on the descending node. Many nongeosynchronous weather satellites are sun synchronous.

sunburn—An inflammation and reddening of the outer skin tissues by extended and unprotected exposure to intense sunlight, especially rich in **UVB** radiation.

sunrise—(1) The phenomenon of the sun's daily appearance on the eastern horizon as a result of the earth's rotation. (2) Contraction for "time of sunrise," defined by the National Weather Service (NWS) as that instant when the upper limb of the sun appears on the sea level horizon; because of **atmospheric refraction** and the size of the solar disk, the center of the disk is 50 minutes of arc below the astronomical horizon at this time.

sunset—(1) The phenomenon of the sun's daily disappearance below the western horizon as a result of the earth's rotation. (2) Contraction for the "time of sunset" defined by the National Weather Service (NWS) practice as that instant when the upper limb of the sun just disappears below the sea level horizon; the effects of **atmospheric refraction** and the semidiameter of the sun are included in this computation.

sunshine—Bright, direct radiation from the sun, as opposed to the shading of a location by clouds or by other obstructions. The duration of bright sunshine measured by a **sunshine recorder** is recorded, along with the **percentage of possible sunshine**, for climatological purposes.

sunshine recorder—An instrument designed to record the duration of bright **sunshine** without regard to intensity at a given location. Most of the sunshine recorders used now in the United States are photovoltaic cells, while others burn strips on specially treated paper.

sunspot—A relatively dark area on the surface of the sun, consisting of a dark central *umbra* surrounded by a *penumbra*, which is intermediate in **brightness** between the umbra and the surrounding **photosphere**.

sunspot cycle—A quasi-periodic variation in the number and area of **sunspots** as given by the relative sunspot number, with an average length of 11.1 years, but varying between about 7 and 17 years. This number rises from a minimum of 0–10 to a maximum of 50–140 about four years later and then declines more slowly. See **solar cycle**.

superadiabatic lapse rate—An **environmental lapse rate** of temperature greater than **the dry-adiabatic lapse rate**, such that **potential temperature** decreases with height. The layer with a superadiabatic lapse rate is **absolutely unstable**.

supercell—Persistent, single, intense **updraft** (usually rotating) and **downdraft** coexisting in a **thunderstorm** in a quasi-steady state rather than in the more usual state of an assemblage of convective cells each of which has a relatively short life; often produces severe weather including **hail** and **tornadoes**.

supercooling—The reduction of temperature of any liquid below the **melting point** of that substance's solid phase; that is, cooling beyond its *nominal* **freezing point** while remaining a liquid.

supergeostrophic—Any **wind speed** greater than the hypothetical **geostrophic wind** speed calculated for that **latitude** and horizontal **pressure gradient**.

superior air—An exceptionally dry mass of air formed by **subsidence** and usually found aloft but occasionally reaching the earth's surface during extreme subsidence processes. It is often found above tropical maritime air, bounded by the **trade wind inversion**.

superior mirage—A spurious image of an object formed above its true position by abnormal **refraction** conditions. Opposite to an **inferior mirage**. Superior mirages occur when the temperature **lapse rate** near the earth's surface is less than its normal value or, especially, when the temperature actually increases with altitude (in an inversion). Usually created over cold surfaces such as ice fields or water bodies.

supersaturation—In meteorology, the condition existing in a given portion of the atmosphere (or other space) when the **relative humidity** is greater than 100%, that is, when it contains more water vapor than is needed to produce **saturation** with respect to a plane surface of pure water or pure ice.

supertyphoon—A classification used by the Joint Typhoon Warning Center (JTWC) to describe a **typhoon** that has sustained near-surface wind speeds of 130 knots or greater.

supplementary observation—Meteorological observation made, in addition to those fixed for a scheduled time, to meet special requirements, such as for **hurricane** forecasting. Contrast with **record observation**.

surf—The sea surface wave activity between the outermost line of **breakers** and the **shoreline**.

surf beat—Irregular oscillations of water level near shore, associated with groups of high **breakers**.

surf zone (also known as *breaker zone*)—The area between the outermost limit of **breakers** and the extent of wave **splash** on the **shoreline**.

surface—(1) The **interface** between the atmosphere and its submedium (land or ocean). (2) For observational purposes, the surface is the horizontal plane located at the observer's elevation, ranging from the height of the **instrument shelter** to **anemometer level**; see **surface weather observations**. (3) A reasonably horizontal continuum, such as an **isobaric surface**, used for reference and analysis purposes.

surface boundary layer—That thin layer of air adjacent to the earth's surface, extending up to the so-called **anemometer level** (the base of the **Ekman layer**). Within this layer the wind distribution is determined largely by the vertical **temperature gradient** and the nature and contours of the underlying surface; shearing **stresses** are approximately constant.

surface chart (also known as ***surface map*** or *surface synoptic chart*)—An analyzed **synoptic chart** of **surface weather observations**. Essentially, a surface chart shows the distribution of **sea level pressure** (therefore, the positions of highs, lows, ridges, and troughs with **isobars**) and the location and nature of fronts and air masses. Often added to this are symbols of occurring **weather** phenomena, analysis of **pressure tendency** (**isallobars**), indications of the movement of pressure systems and fronts, and perhaps others depending upon the intended use of the chart.

surface detention—Water in temporary storage as a thin sheet over the soil surface during the occurrence of **overland flow**.

surface flow—Same as **overland flow**.

surface friction—The **drag** or skin friction of the earth on the atmosphere.

surface inversion—A **temperature inversion** based at the earth's surface; that is, an increase of temperature with altitude beginning at the ground level. See **radiation inversion**.

surface map—Same as **surface chart**.

surface observation—See **surface weather observation**.

surface pressure—In meteorology, the **atmospheric pressure** at a given location on the earth's surface. This expression is applied loosely and about equal to the more specific terms: **station pressure** and **sea level pressure**.

surface radiation budget—Distribution of the **short-** and **longwave radiation** components at the earth's surface; the radiation absorbed by the surface (the **net solar radiation** and downwelling **atmospheric radiation**) and the **terrestrial radiation** emitted by it.

surface retention—Same as **surface storage**.

surface roughness—Irregularities in the surface topography, to include vegetation and buildings, which make the near-surface airflow turbulent and increase the surface **wind stress**.

surface runoff—The water that reaches streams (ranging from the large permanent streams to the tiny rills and rivulets that carry water only during rains) by traveling over the surface of the soil. Thus, surface runoff takes place only over the relatively short distance to the nearest minor channel.

surface storage (also known as ***surface retention***)—The part of precipitation retained temporarily at the ground surface as interception or **depression storage** so that it does not appear as infiltration or **surface runoff** either during the rainfall period or shortly thereafter.

surface temperature—(1) In meteorology, the temperature of the ambient air near the surface of the earth, almost invariably determined by a **thermometer** in an **instrument shelter**. (2) In oceanography, the temperature of the layer of **seawater** nearest the atmosphere. It is generally determined either as **bucket temperature** or **injection temperature**; however, some radiation thermometers are used to measure the "skin temperature" of a layer that is a fraction of a millimeter thick.

surface tension—A phenomenon peculiar to the surface of liquids, caused by a strong attraction toward the interior of the liquid acting on the liquid molecules in or near the surface in such a way to reduce the surface area; produces capillarity and **curvature effects**. Quantitatively, expressed in units of surface energy per unit area.

surface thermometer—A water-temperature thermometer provided with an insulated container around the bulb. The instrument is lowered into the sea on a line until it has had time to reach the temperature of the surface water, then withdrawn and read. The insulated water surrounding the bulb preserves the temperature reading and is available as a **salinity** sample.

surface visibility—Observations of **horizontal visibility** made at an eye level of approximately 6 feet (2 meters) above the ground, using either human evaluation or instrument measurement. This level forms the basis for defining certain **obstructions to vision** (e.g., **fog**, **dust**).

surface water—Water that flows over or is stored on the ground surface. Contrast with **groundwater**.

surface wave—(1) A **gravity wave** formed on the free surface of a **fluid**. (2) In oceanography, sometimes used synonymously with **deep-water wave**.

surface weather observation—An evaluation of the state of the atmosphere as observed from a point at the surface of the earth, as opposed to an **upper-air observation**. This term is applied mainly to **synoptic weather observations**, which are taken for preparing **surface synoptic charts** or to **aviation weather observations** for flight safety purposes at airfields.

surface wind—The wind measured at a surface observing station. This wind is customarily measured by an **anemometer** and **wind vane** at **anemometer level**, usually 10 meters (30 feet) above the ground to minimize the distorting effects of local obstacles and terrain.

surge—(1) In hydrology, a sudden change in **discharge** resulting from the opening or closing of a gate that controls the flow in a channel, or by the sudden introduction of additional water into the channel. (2) Water transported up a beach by breaking waves; see **storm surge**.

surge line—In meteorology, a line along which a **discontinuity** in the **wind speed** occurs. Usually, but not always, the wind speed is strongest upstream from the surge line. Sometimes it is also accompanied by a change in **wind direction**.

sustained winds—In terms of United States observational practice, the winds that have a **wind speed** obtained by averaging the observed values over a one-minute period.

Sverdrup (**Sv**)—A unit of ocean volume **transport** named for Harald Ulrick Sverdrup, an oceanographer; 1 Sverdrup equals one million cubic meters per second.

swash—Intermittent landward flow of water across a beach following the breaking of a wave.

swell—Long-crested ocean waves that have been created by the wind and have traveled out of their **fetch**. Swell usually has flatter crests, longer period, and is more regular than wind-driven waves ("**sea**") in their fetch.

synoptic—(1) In general, pertaining to or affording an overall view. In meteorology, this term has become somewhat specialized in referring to the use of meteorological data obtained simultaneously over a wide area for the purpose of presenting a comprehensive and nearly instantaneous picture of the state of the atmosphere. Thus, to a meteorologist, "synoptic" takes on the additional connotation of simultaneity. (2) Sometimes used in the context of **synoptic-scale** weather systems, such as **cyclones** or **anticyclones**.

synoptic analysis—The diagnostic study of observational data, including the **analysis** of **synoptic charts**, or the body of techniques so employed.

synoptic chart—In meteorology, any chart or map on which data and analyses are presented that describe the state of the atmosphere over a large area at a given moment in time.

synoptic climatology—The study and analysis of climate in terms of synoptic weather information principally in the form of **synoptic charts**. The information thus obtained gives the **climate** (i.e., average weather) of a given locality in a given synoptic situation, rather than the usual climatic parameters, which represent averages over all synoptic conditions.

synoptic code—In general, any code by which **synoptic weather observations** are communicated. The International Synoptic Code presently used by WMO member nations contains the **synoptic reports** from each selected reporting station as a message containing groups of five digit coded words designed for rapid and unambiguous data transmission.

synoptic forecasting—Any forecast methods based upon analysis of a set and/or series of **synoptic charts**.

synoptic hour—A time identified by the hour (**UTC**, **Universal Coordinated Time**), as determined by international agreement, at which **synoptic weather observations** are made simultaneously throughout the world; typically, the primary synoptic hours are every 6 hours commencing at 00:00 UTC.

synoptic meteorology—The study and analysis of weather information (**synoptic charts**, **synoptic weather observations**) taken concurrently over a wide area to ascertain **synoptic-scale** atmospheric systems.

synoptic model—Any **model** specifying a space distribution of some **weather elements**, typically on a **synoptic scale**. The distribution of clouds, precipitation, wind, temperature, and pressure in the vicinity of a **front** is an example of a synoptic model.

synoptic report—An encoded and transmitted **synoptic weather observation** using the international **synoptic code**.

synoptic scale (sometimes called *cyclonic scale*)—Weather phenomena operating at the continental or oceanic spatial scale; includes migratory **cyclones** and **anticyclones**, **air masses**, and **fronts**. This scale lies between **mesoscale** and **planetary scale**.

synoptic situation—The general state of the atmosphere as described by the major weather features of **synoptic charts**.

synoptic weather observation—A **surface weather observation**, made at periodic times or **synoptic hours** (usually at 3-hourly and 6-hourly intervals specified by the World Meteorological Organization), of **sky cover**, **state of the sky**, **cloud height**, **atmospheric pressure** reduced to sea level, **temperature**, **dewpoint**, **wind speed**, and **direction**, amount of **precipitation**, **hydrometeors**, and special phenomena that prevail at the time of the observation or have been observed since the previous specified observation.

system—(1) An entity; an arrangement of organized parts. In meteorology, it often refers to a weather disturbance like an **anticyclone** or **cyclone**. (2) Also used to represent an entity in or approaching **equilibrium**.

syzygy—The points in the moon's orbit about the earth at which the moon appears to be new or full to an earthbound observer.

T

T–S diagram—Abbreviation for **temperature–salinity diagram**.

tabular iceberg—A flat topped-**iceberg** with a length-to-height ratio greater than 5 to 1 that has broken off from an **ice shelf**.

taiga—The open northern part of the **boreal forest**. It consists of open woodland of coniferous trees growing in a rich floor of lichen (mainly ''reindeer moss'' or ''caribou moss''), and is generally cold and swampy. The taiga lies immediately south of the **tundra**. In spring it is often flooded by water from northward-flowing rivers, the lower reaches of which are still frozen.

tailwind—A **wind** that assists the intended progress of an exposed, moving object, for example, rendering an airborne object's **groundspeed** greater than its **airspeed**. Opposite of a **headwind**. The tailwind component is directed along the **heading**, not the **course**.

target—In **radar meteorology**, any of the many types of objects detected by radar, especially those **hydrometeors** such as **raindrops**, **snowflakes**, and **hail**.

target signal—The reflected or scattered microwave **energy** returned to a **radar** unit by a **target**.

target volume—The volume of that part of a precipitation-type radar **target** from which a **target signal** is received. If the precipitation completely fills the radar beam, the target volume is identical with the **radar volume**.

teeth of the gale—An old nautical term for the direction from which the wind is blowing (upwind, **windward**). To sail into the ''teeth of the gale'' or into the ''eye of the wind'' is to sail to windward.

teleconnection—A linkage between weather changes occurring in widely separated regions of the globe.

temperate belt—Based on a **climatic classification** scheme, a belt around the earth within which the annual mean temperature is less than 20°C (68°F) and the mean temperature of the warmest month is higher than 10°C (50°F).

temperate climate—Very generally, the **climatic zone** of the ''middle'' latitudes; the variable climates between the extremes of **tropical climate** and **polar climate**.

temperate rainforest—A type of **rainforest** that exists in cool but generally frost-free regions of heavy annual precipitation. It consists mainly of deciduous trees, usually with one dominant species. With the onset of winter, the forest becomes dormant and remains so until spring when it resumes active growth. Temperate rainforests are found principally near the west coasts of southern South America and northern

North America, in New Zealand, and in northern Japan. Contrast with **tropical rainforest**.

Temperate Zone—Either of the two latitudinal zones on the earth's surface that lie between 23° 27' and 66° 33' N and S (the North Temperate Zone and South Temperate Zone, respectively). These zones are part of one of the **earliest climatic classification** schemes based upon solar illumination geometry, located between the **Torrid** and **Frigid Zones**.

temperature—A measure of the average **kinetic energy** of the individual atoms or molecules composing a substance, a quantity measured by a **thermometer**, specifically with reference to a scale based on defined fiducial points—usually of a water substance, to include the **ice** and **steam points**.

temperature anomaly—(1) Departure of the temperature from its long-term mean value. (2) Difference between the temperature at a specific place and the mean value of the temperature averaged over the circle of latitude through the place.

temperature correction—(1) The correction applied to a **mercury barometer** to adjust the scale reading made at the temperature of the barometer to standard temperature; this correction compensates for the variation in length of the mercury column (**density correction**) and metal scale with temperature. (2) The correction applied to an instrument to account for the effect of temperature upon its response characteristics.

temperature dewpoint spread—Same as **dewpoint depression**; the difference in degrees between the air temperature and the **dewpoint**.

temperature gradient—Change in temperature with distance.

temperature inversion—A layer in which air temperature increases with altitude, representing an "**inversion**" of the typical temperature decrease with height in the **troposphere**. Examples include **radiation**, **frontal,** and **subsidence inversions**.

temperature range—Difference between the **maximum** and **minimum temperatures**, or between the highest and lowest mean temperatures, during a specific time interval.

temperature–humidity index (THI) (also known as *comfort index*)—A measure of human comfort under various combinations of temperature and relative humidity. Compare with the **apparent temperature** and the **heat index**.

temperature–salinity diagram (T–S diagram)—A nomograph used by oceanographers with temperature as ordinate, salinity as abscissa, and water density as a family of curves; the points observed at a single oceanographic station are plotted joined by a curve (the *T–S curve*). Such a plot can be used to identify **water masses** and the stability of the water column with the aid of the lines of constant density.

terminal forecast—An **aviation weather forecast** for one or more specified air terminals. Contrast with **route forecast**.

terminal velocity—The particular falling speed, for any given object moving through a **fluid** medium of specified physical properties, at which the **drag** forces and **buoyant forces** exerted by the fluid on the object just equal the gravitational force acting on the object; after which it falls at constant speed, unless it moves into air layers of different physical properties. See **fall speed**.

terminator—The distinct boundary between the illuminated (daylight) and dark (night) areas of the full disk image of the earth as seen on a **visible satellite image**. Very tall **cumulonimbus** cloud towers are often highlighted in the vicinity of the terminator.

terrestrial radiation—The total **longwave** (**infrared**) **radiation** emitted by the earth's surface; sometimes includes the contribution from its atmosphere (**atmospheric radiation**).

terrestrial scintillation (also known as *shimmer*)—Generic term for **scintillation** phenomena observed in light that reaches the eye from sources lying within the earth's atmosphere; to be differentiated from **astronomical scintillation**, which is observed in light from extraterrestrial sources such as stars.

tertiary circulation—Generally, small, localized atmospheric circulations superimposed upon **primary** and **secondary circulations**; they are represented by such phenomena as **local winds**, **thunderstorms**, and **tornadoes**. See **mesoscale**.

thaw—(1) To free something from the binding action of ice by warming it to a temperature above the melting point of ice. (2) A warm spell when ice and snow melt; as a "**January thaw**."

theodolite—An instrument used to observe the direction of an object in space (e.g., **pilot balloon**) from the observer by the simultaneous determination of its **azimuth** and **elevation angles**.

thermal—A small-scale, rising current of heated air produced above a surface that is warmer and, hence, more buoyant than its immediate surroundings.

thermal capacity—Same as **heat capacity**.

thermal climate—Climate as defined by temperature.

thermal energy—Same as **heat**.

thermal equator—Same as **heat equator**.

thermal high—An **anticyclone** resulting from the cooling of air by a cold underlying surface, and remaining relatively stationary over the cold surface; the atmospheric pressure increases as air subsides during the cooling process.

thermal inertia—Resistance to a change in temperature.

thermal instability—The instability resulting in **free convection** in a **fluid** heated at a boundary.

thermal low—An area of relatively low atmospheric pressure (**cyclone**) due to high temperatures caused by intensive heating at the earth's surface. Thermal lows are common to the continental subtropics in summer; they remain stationary over the area that produces them; their cyclonic circulation is generally weak and shallow; they are nonfrontal. Also called a **heat low**.

thermal pollution—The increase in air or water temperature as a result of the discharge of waste heat from anthropogenic sources, such as power plants; contributes to the urban heat island.

thermal radiation—The **electromagnetic radiation** emitted by any substance as the result of the thermal excitation of its molecules. While this term would apply to any wavelength region, thermal radiation usually refers to the region of the **middle infrared** and **far infrared** regions of spectrum ranging in wavelength from about 3 to 25 micrometers.

thermal tide—A variation in **atmospheric pressure** due to the diurnal differential heating of the atmosphere by the sun; so-called in analogy to the conventional gravitational **tide**.

thermal wind—A term used to describe the vertical **shear** in the **geostrophic wind**; that is, the **vector** difference between the geostrophic **wind vectors** at the base of a layer and the top of the layer. While the term does not apply to a wind, it does describe how the vertical **wind shear** depends upon the horizontal **temperature gradient** in the layer. This wind shear vector (or the thermal wind) parallels the **isotherms** of the mean-layer air temperature, with cold air to the left in the Northern Hemisphere; the magnitude of this shear vector is inversely proportional to the isotherm spacing (i.e., a stronger horizontal temperature gradient produces a greater increase in **wind speed** with height).

thermistor—An **electrical-type thermometer** or thermal sensor whose electrical resistance decreases markedly and monotonically as the temperature increases. Used in electrical thermometers, especially on **radiosondes**.

thermocline—A vertical **temperature gradient** in some layer in a thermally stratified body of water that is appreciably greater than the gradients above and below it; also a layer in which such a gradient occurs. The principal thermoclines in the ocean are either seasonal, due to heating of the surface water in summer, or permanent. In a thermally stratified lake, the thermocline separates the **epilimnion** from the **hypolimnion**.

thermocouple—A thermal sensor utilizing the *thermoelectric effect*, which is the electrical voltage developed in a closed loop when a temperature difference exists between the junctions of the two dissimilar metallic conductors. By having one junction at a known reference temperature, the temperature of the other junction can be determined by measuring the voltage developed in the circuit. Thermocouples are often used in measuring **soil temperatures**.

thermodynamic diagram—Any chart or graph representing values of pressure, density, temperature, water vapor, or functions thereof, such that the **equation of state**, the **Clausius–Clapeyron equation**, and the **first law of thermodynamics** for **adiabatic** and **saturation** (or **pseudo-**) **adiabatic** processes are satisfied. Commonly used for plotting **radiosonde observations**. The **Stüve diagram** and **skew *T*-log*p* diagram** are among the most commonly used.

thermodynamics—The study of the relationships between heat and mechanical energy, especially the heat exchange and the **conservation of energy**. In meteorology, emphasis is placed upon the **change of state** of the gaseous atmosphere as a result of various physical processes.

thermograph—A self-recording **thermometer**.

thermohaline circulation—Circulation in a large water body caused by changes in water density brought about by the combined effect of variations in **temperature** and **salinity**.

thermometer—An instrument used for measuring **temperature** by incorporating a thermal sensor that utilizes the variation of the physical properties of substances according to their thermal states. Examples include **liquid-in-glass**, **deformation-type**, **electrical type**, and **radiation thermometers**.

thermosphere—The atmospheric shell extending from the temperature minimum at the **mesopause** (the top of the **mesosphere**) to outer space. It is a region of more or less steadily increasing temperature with altitude, starting at an altitude of about 80 kilometers (50 miles). The thermosphere includes the **exosphere** and most or all of the ionosphere.

third law of thermodynamics—The statement that every substance has a finite positive **entropy**, and the entropy of a crystalline substance is theoretically zero at the temperature of **absolute zero**.

thunder—The sharp or rumbling sound emitted by rapidly expanding gases along the channel of a **lightning discharge**; an **electrometeor**. The intensity and type of sound depends upon atmospheric conditions and the distance between the lightning discharge and the observer.

thunderbolt—In mythology, a **lightning flash** accompanied by a material ''bolt'' or ''dart''; this is the legendary cause of the damage done by lightning. It is still used as a popular term for a **cloud-to-ground lightning discharge** accompanied by thunder.

thundercloud—A convenient and often used term for the cloud mass of a **thunderstorm**; that is, a **cumulonimbus**.

thunderhead—A popular term for the **anvil** (**incus**) of a **cumulonimbus** cloud; or, less appropriately, the upper portion of a swelling cumulus, or the entire cumulonimbus.

thundershower—A colloquial term used for a **thunderstorm**.

thundersnow—A colloquial term used for a **thunderstorm** with snow, that is, any convective snowfall or **snow burst** accompanied by lightning and thunder.

thunderstorm—In general, a local storm invariably produced by a **cumulonimbus** cloud, and always accompanied by **lightning** and **thunder**, usually with strong wind **gusts**, heavy rain, and sometimes with **hail**. Classification is often made according to the forcing mechanism to include **advective thunderstorms, air mass thunderstorms**, and **frontal thunderstorms**.

thunderstorm cell—The convection cell of a **cumulonimbus** cloud.

thunderstorm charge separation—The process by which the large **electric fields** found within **thunderclouds** are generated; the process by which particles bearing opposite electrical charge are given those charges and transported to different regions of the active cloud.

thunderstorm day—An observational day during which **thunder** is heard at the station; precipitation need not occur.

thunderstorm outflow—Outward flow of air from under a **thunderstorm** resulting when a **downdraft**, usually relatively cold, is deflected laterally as it approaches ground level.

tidal bore—A rapidly advancing wall-like wave of water caused by a very rapid rise of the tide (or an advancing **tsunami** usually within a long, shallow, and narrowing **estuary**, with a large **tidal range**. Essentially the same as a **hydraulic jump**.

tidal component—Same as **partial tide**.

tidal current—The alternating horizontal movement of water associated with the vertical rise and fall of the **tide** caused by luni-solar **tide-producing forces**. In relatively open positions, the direction of tidal currents rotates continuously through 360° diurnally or semidiurnally, as indicated on a **current chart**. In coastal regions, the nature of tidal currents will be determined by local topography as well.

tidal day—Same as **lunar day** (1).

tidal excursion—The net horizontal distance over which a water particle moves during one **tidal cycle** of **flood** and **ebb tide**. The distances traversed during ebb and flood are rarely equal in nature, since a layered circulation usually occurs in an **estuary**, with a net surface flow in one direction compensated by an opposite flow at depth.

tidal prism—The volume of water that moves in or out of a harbor or other basin during each **tidal cycle**. It is usually computed as the product of the **tidal range** and the area of the basin at midtide level.

tidal range—The difference in height between consecutive **high water** and **low water**; twice the **tide amplitude**.

tidal wave—(1) Astronomically, the periodic variations of sea level caused by the luni-solar gravitational attractions. (2) In popular (but not necessarily correct) usage, any unusually high (and therefore destructive) water level along a shore. It usually refers to either a **storm surge** or **tsunami**.

tidal wind—A very light breeze that occurs in calm weather in inlets where the **tide** sets strongly; it blows onshore with rising tide and offshore with ebbing tide.

tide—The periodic rising and falling of the earth's oceans and atmosphere resulting from the **tide-producing forces** of the moon and sun acting upon the rotating earth. This disturbance actually propagates as a wave through the atmosphere and along the surface of the waters of the earth.

tide amplitude—One-half of the difference in height between consecutive **high water** and **low water**; hence, half of the **tidal range**.

tide curve—A graphical portrayal of the rise and fall of the **tide** as a function of time; traditionally, time in hours (of days) is plotted on the abscissa and height of the tide as the ordinate.

tide cycle—A complete set of tidal variations or characteristics covering one period, such as a **tidal day**, a **lunar month**, or a **Metonic cycle**.

tide gauge—A device for measuring the height of **tide**. It may be simply a graduated staff in a sheltered location where visual observations can be made at any desired time; or it may consist of an elaborate recording instrument (sometimes called *marigraph*) making a continuous graphic record of tide height against time.

tide tables—Annual tabulations of daily predictions of the times and heights of **high water** and **low water** at various places.

tide-producing forces—The interactions associated with the gravitational attraction between two astronomical bodies and the **centripetal force**.

tilt—In **synoptic meteorology**, the **inclination** to the vertical of a significant feature in a given meteorological **field**, such as wind, pressure, temperature, or moisture fields. For example, troughs in the westerlies usually display a westward tilt with altitude in the lower and middle **troposphere;** that is, the center of the **trough** on each chart at successively higher altitudes would be displaced farther toward the west.

timber line (also known as ***tree line***)—(1) In mountainous regions, the line above which climatic conditions do not allow the upright growth of trees. (2) The poleward limit of tree growth.

time constant—Generally, the time required for an instrument to indicate a given percentage of the final reading resulting from an input signal; used to describe instrument response.

time lag—The total time between the application of a signal to an instrument and the full indication of that signal within the uncertainty of the instrument.

time series—The values of a variable generated successively in time.

tipping-bucket rain gauge—A self-recording instrument that collects **rainfall** in increments of 0.25 millimeter (0.01 inch) by containers that alternately fill, tip, and empty; an example of a **rain-intensity gauge**, since rain intensity can be determined from the number of tipping cycles per time increment recorded on the chart.

topography—Generally, the disposition of the major natural and human-built physical features of the earth's surface, such as forests, rivers, highways, bridges, etc. These topographic features are of the type that would be entered on a map, to include **contours** of **elevation**, although the term is often used to denote elevation characteristics (particularly **orographic** features) alone.

tornado—A small mass of air (**whirlwind**) that spins rapidly about an almost vertical axis and forms a **funnel cloud** that contacts the ground; appears as a pendant from a **cumulonimbus** cloud and is potentially the most destructive of all weather systems. Tornadoes have diameters ranging from ten meters to over several hundred meters.

tornado alley—The area of the United States in which tornadoes are most frequent. It encompasses the great lowland areas of the Mississippi, the Ohio, and lower Missouri River valleys. Although no state is entirely free of tornadoes, they are most frequent in the Plains area between the Rocky Mountains and Appalachians.

tornado echo—A type of **radar echo** of the precipitation which appears as a hook in the southwest sector of a **thunderstorm** and often appears in conjunction with a **tornado**. Same as **hook echo**.

tornado outbreak—Formation of a large number (typically, six or more) of **tornadoes**, either in groups or as individuals, within a 24- to 48-hour period over a particular area; usually these tornadoes are spawned by the same general weather system.

tornado vortex signature (TVS)—A characteristic **Doppler radar** velocity image showing a small region of rapidly changing **wind direction** within a **mesocyclone**; indicative of a tornadic circulation.

tornado warning—Issued by the National Weather Service to warn the public that a **tornado** has been sighted by storm spotters or has been indicated by **radar**. These warnings are issued with information concerning where the tornado is presently located and what communities are in the anticipated path of the tornado. The public is expected to heed these warnings and take appropriate action.

tornado watch—Issued by the National Weather Service to alert the public that conditions are favorable for the development of **tornadoes** in and close to the watch area, which is delineated in the tornado watch statement. These watches are issued with information concerning the watch area and the length of time they are in effect; watches are usually in effect for several hours.

Torrid Zone—The zone of the earth's surface between the **Tropics of Cancer** and **Capricorn**. This is one of the three subdivisions of the earliest **climatic classification** scheme based upon solar illumination geometry; the other two are the **Temperate Zone** and the **Frigid Zone**.

Total Ozone Mapping Spectrometer (TOMS)—Flown initially on NASA's *Nimbus-7* satellite, this instrument's primary function is high-resolution global mapping of total **ozone** on a daily basis. TOMS maps the total amount of ozone between the ground and the top of the atmosphere and provides maps of the **antarctic ozone hole**.

Tower of the Winds—An octagonal, marble building in Athens erected not later than 35 B.C. and still standing. The sides face the points of the Athenian compass and carry a frieze of male personifications of the winds from those directions: *Boreas* (N), *Kaikias* (NE), *Apheliotes* (E), *Euros* (SE), *Notos* (S), *Lips* (SW), *Zephyros* (W), and *Skiron* (NW). These figures are reproduced on the tower of the Radcliffe Observatory, Oxford, England, and in the Library of the Blue Hill Observatory near Boston, Massachusetts. The Tower was not a meteorological observatory, though it originally carried a weather vane on the roof, but was built to measure time, the walls bearing sundials, with a water clock inside for use during cloudy weather.

towering cumulus—A descriptive term, used mostly in weather observing, for **cumulus congestus** of great vertical extent.

trace—(1) A precipitation amount of less than 0.25 millimeter (0.01 inch). (2) An amount of cloud less than 1/10 of the **sky** dome. (3) The record made by any self-registering instrument. Thus, one may speak of a barograph trace, a hygrothermograph trace, etc.

trace element—A chemical element present in minute amounts in the atmosphere.

trade wind belt—The general belt of latitude occupied by the **trade winds**.

trade wind cumulus—The characteristic **cumulus** cloud of the trade winds over the tropical oceans in average, undisturbed weather conditions. These clouds are, generally, 1500–2100 meters (5000–7000 feet) thick at peak development and are based at about 610–762 meters (2000–2500 feet) altitude. The individual cloud usually exhibits a blocklike appearance since its vertical growth ends abruptly in the lower stratum of the **trade-wind inversion**. A group of fully grown clouds shows considerable uniformity in size and shape.

trade wind inversion—A characteristic **temperature inversion** usually present in the **trade wind** streams over the eastern portions of the tropical oceans. It is formed by broad-scale **subsidence** of air from high altitudes in the eastern extremities of the **subtropical highs**. While descending, the adiabatically heated current meets the opposition of the low-level, cool maritime air flowing equatorward, enhancing the intensity of the inversion. The inversion forms at the meeting point of these two strata, which flow horizontally in the same direction. Trade wind inversions suppress convection, especially in the eastern portions of ocean basins where a stronger inversion is typically closer to the surface.

trade winds (or *trades*)—The wind system, occupying most of the Tropics, which blows outward from the **subtropical highs** toward the **equatorial trough**; a major component of the planetary-scale circulation of the atmosphere. Surface winds are northeasterly in the Northern Hemisphere and southeasterly in the Southern Hemisphere; hence, they are known as the ***northeast trades*** and ***southeast trades***, respectively.

trades—Common contraction for **trade winds**.

trailing front—A **cold front** extending outward from a primary **cyclone** with a large latitudinal extent along which a **cyclone family** may develop.

trajectory—(1) A continuous line in space tracing the successive positions of a moving particle of air over a given time interval. (2) Line in space tracing the successive positions of the center of a **synoptic system**.

transmissometer—An instrument used to measure **horizontal visibility**, especially at airport runways; the transmission or extinction of a light beam over a fixed distance is measured.

transmittance—A measure of the amount of **radiation** propagated through a given medium; defined as the (dimensionless) ratio of the radiation transmitted through a medium to the total radiation incident on it.

transparency—Property of a medium, which enables a stream of incident **radiation** to pass through it, as opposed to being *absorbed*, *scattered*, or *reflected* by the medium.

transpiration—The process by which water in plants is transferred through leaf pores and converted into water vapor to the atmosphere; a part of **evapotranspiration**.

transport—The process by which a substance or quantity is carried past a fixed point, or across a fixed plane of unit area; also known as **flux**. In meteorology and oceanography, such quantities are heat, momentum, mass, dissolved impurities, suspended particles, etc.

trapping—A pollution **plume** that is kept from dispersing upward because of weakly stable atmospheric conditions extending to above the plume; a **temperature inversion** may cap the layer below. **Mixing** of the plume downward is not as extreme as in the **fumigation case**.

tree line (also known as ***timber line***)—(1) The poleward limit of tree growth; the botanical boundary between tundra and boreal forest. The arctic tree line has been studied extensively, but in the Southern Hemisphere, a ''tree line'' can be inferred only by comparison of vegetation on islands in the southern oceans. (2) In mountainous regions, the line above which climatic conditions do not allow tree growth.

tree ring chronology—See **dendrochronology**.

triple point—(1) The thermodynamic state at which three physical phases of a substance exist in **equilibrium**. (2) A junction point within the Tropics of three distinct **air masses**, considered to be an ideal point of origin for a **tropical cyclone**. (3) The **point of occlusion** of an **extratropical cyclone** where the **cold**, **warm**, and **occluded fronts** meet; a preferred site for **cyclogenesis**.

Tropic of Cancer—The northern parallel of maximum **solar declination**, approximately 23° 27'N, at the northern boundary of the **Torrid Zone**.

Tropic of Capricorn—The southern parallel of maximum **solar declination**, approximately 23° 27'S, marking the southern boundary of the **Torrid Zone**.

tropic tide—An ocean **tide** occurring when the moon is near maximum **declination**; the **diurnal inequality** is then at a maximum.

tropical air—An **airmass** type whose characteristics are developed over low latitudes. Contrast with **polar air**. Maritime tropical air (*mT*), the principal type, is produced over the tropical and subtropical seas. It is very warm and humid, and is frequently carried poleward on the western flanks of the **subtropical highs**. Continental tropical air (*cT*) is produced over subtropical arid regions and is hot and very dry.

tropical climate—In general, a **climatic zone** with a climate typical of equatorial and tropical regions; that is, one with continually high temperatures and with considerable precipitation, at least during part of the year.

tropical cyclone—Generic term for a nonfrontal **synoptic-scale cyclone** originating over tropical or subtropical waters with organized convection and definite **cyclonic** surface wind circulation; contrast with **extratropical cyclone**. In early stages, tropical cyclones move from east to west in the broad zone of prevailing easterlies. Additional classification of tropical cyclones is made depending upon organization and **wind speed**; according to international agreement these include a **tropical disturbance**, **tropical depression**, **tropical storm**, and **hurricane** (or **typhoon**).

tropical depression—A **tropical cyclone** in which the maximum 1-minute sustained surface wind is 33 knots (38 mph) or less.

tropical disturbance—A discrete system of apparently organized convection—originating in the Tropics or subtropics—having a nonfrontal migratory character and maintaining its identity for 24 hours or more. It may be associated with a detectable perturbation of the wind field. As such, it is the basic generic designation that, in successive stages of intensification, may be subsequently classified as a **tropical wave**, **tropical depression**, **tropical storm**, or **hurricane/typhoon**.

tropical easterlies—A term applied to the **trade winds** when they are shallow and exhibit a strong vertical **wind shear**. With this structure, at about 1500 meters (4900 feet) the easterlies give way to the upper westerlies (**antitrades**), which are sufficiently strong and deep to govern the course of cloudiness and weather.

tropical front—Same as **intertropical front**.

tropical meteorology—The study of the tropical atmosphere and weather systems of the **Tropics**.

tropical rainforest—A type of **rainforest** that exists in tropical regions where precipitation is heavy, generally more than 250 centimeters (98 inches) per year. It consists mainly of a wide variety of lofty trees, which carry a profusion of parasitic or climbing plants, and, in some portions, a ''jungle'' of dense undergrowth near the ground. For lack of marked climatic seasons, growth proceeds throughout the year.

tropical rainforest climate—In general, the climate that maintains **tropical rainforest** vegetation; that is, a climate of unbroken warmth, high humidity, and heavy annual precipitation.

tropical savanna climate—In general, the type of climate that produces the vegetation of the tropical and subtropical **savanna**; thus, a climate with a winter **dry season**, a relatively short but heavy summer **rainy season**, and high temperatures year-round.

tropical storm—A **tropical cyclone** of maximum **wind speed** of 34–63 knots (39–74 mph) according to United States classification. The international classification defines a tropical storm with wind speeds between 34 and 47 knots and a *severe tropical storm* between 48 and 63 knots.

tropical storm warning—Issued by the National Weather Service to warn the public of imminent **tropical storm** conditions (sustained winds within the range 34–63 knots), which are expected in a specified coastal area within 24 hours or less.

tropical storm watch—Issued by the National Weather Service to alert the public in a coastal area when a threat of **tropical storm** conditions (sustained winds within the range 34 to 63 knots) is predicted within 36 hours.

tropical wave—Same as **easterly wave**. A migratory wavelike disturbance of the **tropical easterlies** usually moving westward.

tropical wet and dry climate—Same as **tropical savanna climate**.

tropical wet climate—Same as **tropical rainforest climate**.

Tropics—(1) Same as **Torrid Zone**. (2) Any portion of the earth characterized by a **tropical climate**.

tropopause—The boundary between the **troposphere** and **stratosphere**, usually characterized by an abrupt change of the temperature **lapse rate** to less than 2°C per kilometer for an extended depth (usually 2 kilometers). The change is in the direction of increased atmospheric stability from regions below to regions above the tropopause. Its average altitude varies from about 16 kilometers (10 miles) in the Tropics to about 6 kilometers (4 miles) in polar regions. In polar regions in winter locating the tropopause is often difficult or impossible, since under some conditions no abrupt change in lapse rate can be detected at any altitude.

tropopause inversion—The decrease in the **lapse rate** of temperature encountered at the level of the **tropopause**, culminating in an increase in temperature with height in the lower **stratosphere**, forming a **temperature inversion**.

troposphere—That portion of the atmosphere from the earth's surface to the **tropopause**; that is, the lowest 6–16 kilometers (4–10 miles) of the atmosphere. The troposphere is characterized by decreasing temperature with altitude, appreciable vertical wind motion, appreciable water vapor, and **weather**. Dynamically, the troposphere can be divided into the following layers: **surface boundary layer**, **Ekman layer**, and **free atmosphere**.

trough—(1) Also known as *wave trough,* representing the lowest part of a **wave**. Contrast with **crest**. (2) In meteorology, an elongated area of relatively low **atmospheric pressure** at a given level; usually an elongated area of **cyclonic** curvature of the wind **field**. Opposite of a **ridge**. The axis of a trough is the *trough line*.

trough aloft—Same as **upper-level trough**.

true altitude (also called ***corrected altitude***)—The true vertical distance above **mean sea level**. Contrast with **absolute altitude**.

true north—The direction from any point on the earth's surface toward the geographic North Pole; the northerly direction along any projection of the earth's rotational axis upon the earth's surface, for example, along a **meridian** of **longitude**. Except for some navigational practices which use **magnetic north**, true north is the universal 0° (or 360°) mapping reference. True north differs from magnetic north by the magnetic **declination** (2) at that geographic point.

tsunami (also known as *seismic sea wave*)—An ocean wave produced by a submarine earthquake, landslide, or volcanic eruption. These waves may reach enormous dimensions and have sufficient energy to travel across entire oceans. They proceed as ordinary **gravity waves** with a period between 15 and 60 minutes.

tsunami watch/warning—Issued by the National Weather Service to either alert or warn residents in regions along the Pacific Ocean that an impending **tsunami** may cause damage to low lying regions. The type of bulletin is based on the magnitude and the location of the underwater geological event. The content and format is similar to **coastal flood watches** and **warnings**.

Tsushima Current—A warm, northward-flowing ocean current in the North Pacific following the western coast of Japan; a part of the **Kuroshio System**. The Tsushima Current branches off on the left-hand side of the **Kuroshio** flowing north into the Japan Sea.

tundra—Treeless plains that lie poleward of the **tree line**. The plants thereon are sedges, mosses, lichens, and a few small shrubs. It is mostly underlain by **permafrost**, so that drainage is bad and the soil may be saturated for long periods. It does not have a permanent **snow-ice cover**. Also called **arctic desert**.

turbidity—In meteorology, any condition of the atmosphere that reduces its **transparency** to radiation, especially to **visible radiation**. Ordinarily, this condition is applied to a cloud-free portion of the atmosphere that owes its turbidity to air molecules and suspended particles such as **smoke**, **dust**, and **haze**, and to **scintillation** effects.

turbidity current—A **density current** containing a mass of agitated sediment that flows down a submarine canyon as a result of density differences between the denser turbid water and the less dense surrounding water. The current ultimately spills out onto a deep-sea plain.

turbulence—A state of **fluid** flow in which the instantaneous velocities exhibit irregular and apparently random fluctuations so that in practice only statistical properties can be recognized and subjected to analysis; this condition can be superimposed upon mean fluid flow.

turbulent diffusion (also known as *eddy diffusion*)—The **transport** and dispersal of matter or other atmospheric properties, such as heat and momentum, by **eddies** in turbulent flow.

twilight—An intermediate period of **illumination** of the sky before **sunrise** and after **sunset** as a consequence of the multiple scattering of sunlight by constituents of the upper atmosphere. Twilight is often broken into **astronomical**, **civil**, and **nautical twilight**.

twilight arch (also known as ***bright segment***)—A faint glowing purple band that appears above the horizon in the direction of the sun after sunset or before sunrise; a **twilight phenomenon**.

twilight phenomena—Those meteorological optical phenomena that occur during **twilight**, including such effects as the **green flash**, **bright segment** (or **twilight arch**), **purple light**, and **crepuscular rays**.

twister—In the United States, a colloquial term for **tornado**.

typhoon—A severe **tropical cyclone** with sustained surface wind speeds greater than 64 knots (74 mph) in the western North Pacific (west of the International Dateline); equivalent of a **hurricane**. See **supertyphoon**.

typhoon warning—See **hurricane/typhoon warning**.

typhoon watch—See **hurricane/typhoon watch**.

U

Ultra Violet Index (UVI)—A numeric value, ranging from 0 to 15, used by the National Weather Service to describe the anticipated exposure level to **ultraviolet radiation** for a particular location at local solar noon. Levels of exposure are *minimal* (0–2), *low* (3–4), *moderate* (5–6), *high* (7–9), and *very high* (10 and higher). The Index includes the effects of cloud cover on the anticipated ultraviolet exposure.

ultraviolet radiation (UV)—Energetic **electromagnetic radiation** with wavelengths ranging from 1 to 300 nanometers (10 **Ångstroms**–0.3 micrometer) longer than **X-rays** but shorter than **visible light**. Although ultraviolet radiation constitutes only about 5% of the total energy emitted by the sun, it is responsible for producing the temperature structure of the **stratosphere** and **mesosphere**, playing a dominant role in both energy balance and chemical composition. Much of the solar ultraviolet radiation is absorbed in the stratosphere where it is involved in the natural formation and destruction of **ozone**. Ultraviolet radiation can be divided into **UVA**, **UVB**, and **UVC**.

undercurrent—In oceanography, a water current flowing beneath a **surface current** at a different speed or in a different direction.

unit hydrograph—A statistically derived **hydrograph** for a storm of specified duration that results in 1.00 inch of direct runoff within the **drainage area** concerned. It is expressed as a graph or table of **discharge** as a function of time.

Universal Coordinated Time (**UTC**)—The global system of time based on the local **civil time** observed on the **Prime Meridian**, previously called *Greenwich Mean Time* (GMT) or **"Zulu" (Z) time**. A 24-hour format is used.

unlimited ceiling—A **ceiling** condition that exists: (*a*) when the total **sky cover** is less than 0.6; (*b*) when the total transparent sky cover is 0.5 or more; or (*c*) when surface-based **obscuring phenomena** are classed as **partial obscuration** (that is, obscures 0.9 or less of the sky) and no layer aloft is reported as **broken** or **overcast**.

unrestricted visibility—The **visibility** when no **obstructions to vision** exist in sufficient quantity to reduce the visibility to less than 7 miles (11 kilometers).

unsaturated zone—See **zone of aeration**.

unstable air—A layer of air characterized by a vertical **temperature gradient** such that **air parcels** displaced upward (downward) will accelerate upward (downward) and away from their original altitude. Contrast with **stable air**.

unstable motion—Motion in which small disturbances grow. Contrast with **stable motion**.

updraft—Upward-moving current of air of small dimensions.

upper air—In **synoptic meteorology** and **weather observation**, that portion of the atmosphere that is above the lower **troposphere**. No distinct lower limit is set but the term is generally applied to the levels above 850 hectopascals (millibars).

upper atmosphere—The general term applied to the earth's atmosphere above the **mesopause**, or above an altitude of 85 kilometers. See **lower** and **middle atmosphere**.

upper-air chart—Same as **upper-level chart**.

upper-air disturbance—A **cyclonic** flow pattern aloft, often accompanied by clouds; the term is used particularly for such a system that is more strongly developed aloft than near the ground.

upper-air observation—A measurement of atmospheric conditions aloft, above the effective range of what constitutes a **surface weather observation**.

upper-level chart—A **synoptic chart** of meteorological conditions in the **upper air**, almost invariably referring to a standard **constant-pressure chart** made at one of the **mandatory levels**.

upper-level disturbance—Same as **upper-air disturbance**.

upper-level ridge—A pressure **ridge** existing in the **upper air**, especially one that is stronger aloft than near the earth's surface.

upper-level trough—A pressure **trough** existing in the **upper air**, especially one that is more pronounced aloft than near the earth's surface.

upslope fog—A type of **fog** formed when humid, stable air flows upward over rising terrain and is, consequently, adiabatically cooled to or below its original **dewpoint**.

upstream—In the direction from which a **fluid** is flowing. Opposite of **downstream**.

upwelling—(1) The component of radiation (either reflected **solar** or emitted **terrestrial**) directed upward from the earth's surface. (2) The rising of water toward the surface from deep subsurface layers of a water body associated with near-surface **divergence**. The surface contains cold nutrient-rich water. The opposite of **downwelling** (2).

upwind—In the direction from which the wind is blowing. Opposite of **downwind**.

upwind effect—The effect of an orographic barrier in producing **orographic precipitation** at a distance **windward** of the base of the barrier. This phenomenon results because the moist airflow is forced upward before the barrier slope is actually reached.

urban and small stream flood advisory—Issued by the National Weather Service to alert the public to potential flooding conditions that would appear to cause inconvenience (but not life-threatening) to those living in the affected area. These advisories are issued when heavy rain is anticipated, which could cause flooding of streets and other urban low-lying regions, or if small streams are expected to reach bankfull.

urban climate—Climate of cities, which differs from that of the surrounding areas because of the influence of the urban settlement.

urban heat island—The relative warmth of a city compared with surrounding rural areas; **thermal pollution** contributes to this effect.

UTC—The official abbreviation for **Universal Coordinated Time**.

UVA—A portion of solar **ultraviolet radiation** that reaches the earth's surface; in the wavelength range of 320 to 400 nanometers; responsible for tanning of the skin without **sunburn** and some types of skin cancers.

UVB—A biologically effective portion of solar **ultraviolet radiation** that reaches the earth's surface; in the wavelength range of 280 to 320 nanometers; responsible for **sunburns** and skin cancers.

UVC—A portion of solar **ultraviolet radiation** in the wavelength range of 100 to 280 nanometers; this potentially lethal radiation does not reach the earth's surface, because various atmospheric constituents such as oxygen and ozone absorb this radiation.

UVI—Abbreviation for **Ultra Violet Index**

V

valley fog—A fog that develops in a topographic depression or valley due to a combination of nocturnal **radiational cooling** and cold **air drainage**, which cause a pooling of air that is easily cooled to temperatures below the original **dewpoint**. See also **radiation fog**.

valley glacier—A **glacier** that flows down a valley.

valley wind—A localized **anabatic wind** that ascends a mountain valley (up-valley wind) during daylight as a consequence of surface heating; the daytime component of a **mountain and valley wind system**. This wind regime is most prevalent when the region experiences calm, sunny conditions.

Van Allen radiation belts—Doughnut-shaped regions encircling earth and containing high energy electrons and ions trapped in the earth's **magnetic field**. These bands, named for their discoverer, James A. Van Allen, (1914–) a physicist at the University of Iowa, are in equatorial orbits extending from the earth at distances ranging between approximately 6000 and 13 000 kilometers and from 19 000 to 25 000 kilometers.

vapor—Any substance existing in the gaseous state at a temperature lower than that of its **critical point**; that is, a gas cool enough to be liquefied if sufficient pressure were applied to it. In meteorology and the other earth sciences, vapor usually refers to **water vapor**.

vapor pressure—The **partial pressure** exerted by the molecules of a given **vapor**, which in meteorology is usually **water vapor**. For a pure, confined vapor, it is the vapor's pressure on the walls of its containing vessel; and for a vapor mixed with other vapors or gases, it is that vapor's contribution to the total pressure (i.e., its partial pressure) in accordance with **Dalton's law** of partial pressures.

vaporization—Often used synonymously with **evaporation**.

variable ceiling—In United States weather observing practice, a condition in which the height of the base of a **broken** or **overcast** cloud layer (**ceiling**) rapidly increases and decreases while the ceiling observation is being made. The average of the observed values is used as the reported ceiling, and it is reported only for ceilings of less than 3000 feet.

variable visibility—In United States weather observing practice, a condition in which the **prevailing visibility** fluctuates rapidly while the observation is being made. The average of the observed values is used as the reported visibility, and it is reported only for visibilities of less than 3 miles (5 kilometers).

variable wind direction—According to United States weather observing practice, a condition in which the **wind direction** of the **surface wind** is observed to fluctuate by 60° or more during the 10-minute observation interval.

variables of state—The minimum numbers of descriptors of the physical **state** (2) of a thermodynamic system; for a gas these would include the absolute temperature, pressure, and mass density of the gas, such as air.

varves—The annual layers of sediment deposited in lakes and fjords by **meltwater** from glaciers. Each layer consists of two parts deposited at different seasons and differing in color and texture so that the layers can be measured and counted. If the series were complete, the number of layers gives the date on which the ground was vacated by the retreating ice.

vector—A quantity that needs both a direction and magnitude for complete specification. Examples include **velocity**, **force**. Contrast with **scalar** quantities.

veering—A change in **wind direction** in a clockwise sense (e.g., south to southwest to west) in the Northern Hemisphere of the earth. Opposite of **backing**.

velocity—The time rate of change of a position **vector**; that is, a change of position expressed as a vector quantity in terms of both *speed* (a magnitude) and *direction*. Units of the magnitude of velocity are length per time, such as meters per second, or miles per hour.

Venturi effect—Local decrease of pressure, local increase of the wind, and the appearance of gusts in certain places when the wind blows through a narrow mountain pass. Named for G.B. Venturi, an Italian physicist.

verification time—The time when the forecast is to be valid, or "verified" by observations.

vernal—Pertaining to **spring**. The corresponding adjectives for **summer**, **fall**, and **winter** are **aestival**, **autumnal**, and **hibernal**.

vernal equinox—The **equinox** at which the sun approaches the Northern Hemisphere, marking the start of astronomical **spring**; the time of this occurrence is approximately 21 March. Contrast with **autumnal equinox**.

vertical beam radar—Radar fixed to probe only vertically and giving the distribution of **radar echo** intensity with altitude.

vertical profile—Graph showing the variation of a **weather element** (e.g., temperature, dewpoint) with altitude.

vertical stretching—A process in which the ascending vertical motion of air increases with altitude, or descending motion decreases with (increasing) altitude.

vertical velocity—In meteorology, the component of the **velocity vector** along the local vertical; mathematically, the vertical wind component directed upward is positive in a local coordinate system.

vertical visibility—In United States weather observing practice, the distance that an observer can see vertically into a surface-based **obscuring phenomenon** such as **fog**, **rain**, or **snow**. This estimate must be based upon penetration of the beam of a

ceilometer, or less frequently, by **pilot balloon** ascension or ceiling light. The height is reported as **ceiling height**.

vertically developed clouds—The class (or étage) of clouds based on altitude containing **cumulus** and **cumulonimbus** clouds that may extend upward through more than one of the layers (étages) used to classify **low**, **middle**, and **high clouds**.

VFR weather—In aviation terminology, route or terminal weather conditions that allow operation of aircraft under **visual flight rules** (VFR); **ceilings** are greater than 3000 feet (900 meters) and **visibility** greater than 5 miles (8 kilometers).

violent storm—Wind with a speed between 56 and 63 knots (63–72 mph); on the modern international **Beaufort wind scale** a *Beaufort force 11*, producing widespread damage and unusually high waves.

VIP level—An abbreviation for Video Integrator Processor, representing one of the six levels of **radar reflectivity,** ranging from level 1 (*light* precipitation) to level 6 ("*extreme*" or very heavy rain and hail); these contoured **echo intensity** levels are often color coded.

virga (same as *fallstreaks*)—Wisps or streaks of water or ice particles falling out of a cloud, much of which vaporizes before reaching the earth's surface. Virga is also detected by **weather radar** and reported as *precipitation aloft.*

virtual temperature—In a system of **humid air**, the temperature of **dry air** having the same density and pressure as the **humid air**.

VIS—Same as **visible satellite imagery**.

viscosity—That molecular property of a **fluid** that enables it to support tangential **stresses** for a finite time and thus to resist **deformation** or shearing.

viscous fluid—A **fluid** whose molecular **viscosity** is sufficiently large to make the viscous forces a significant part of the total force field in the fluid.

visibility—The greatest distance from an observer that a permanent object of known characteristics can be seen and identified by unaided, normal eyes. Most often refers to **horizontal visibility**. In United States weather observing practice, the greatest distance in a given direction at which it is just possible to see and identify with the unaided eye (*a*) in the daytime, a prominent dark object against the sky at the horizon, and (*b*) at night, a known, preferably unfocused, moderately intense light source. After visibilities have been determined around the entire horizon circle, they are resolved into a single value of prevailing visibility for reporting purposes. Now most commonly determined by light-scattering techniques where decreasing visibility is correlated with increased scattering. See **surface visibility**, **vertical visibility**, and **oblique visibility**.

visible radiation—That part of the **electromagnetic radiation** spectrum lying within the wavelength interval to which the normal human eye is sensitive, the spectral interval from approximately 0.39 to 0.76 micrometer, or between **ultraviolet** and **infrared**. This region corresponds to the peak in the solar radiation emission curve.

visible satellite imagery (VIS)—Images from **radiometers** onboard a satellite that sense **visible solar radiation** (typically, 0.55–0.99 micrometer) reflected or back-scattered from surfaces in the earth–atmosphere system. Contrast with **infrared satellite image**.

visual range—The distance, under daylight conditions, at which the apparent contrast between a specified type of target and its background becomes just equal to the threshold contrast of an observer. See **runway visual range**.

volcanic ash—Fine particles of rock powder and other material ejected from a volcano and which can be suspended in the stratosphere for long time intervals. It may represent an **obstruction to vision** or **obscuration**, hence, reported as a **lithometeor**. Since the ash represents a potential hazard to aircraft operation, current United States observation practice requires reporting of this lithometeor when present, even if **visibility** were not reduced.

vortex—Any circular flow possessing **vorticity**; more often the term refers to a flow with closed **streamlines**. See **eddy** or **whirlpool**.

vorticity—Measure of the rotational spin about an axis at some point within a **fluid**. Although any orientation of the spin axis is possible, in most meteorological applications, the vertical vorticity component describing the rotation about a vertical axis is used. See **relative** and **absolute vorticity**.

W

wake—The region of turbulence immediately to the rear of a solid body in motion relative to a **fluid**.

Walker Circulation—Direct zonal tropical circulation, thermally driven, in which air rises over the warm western Pacific Ocean and sinks over the cold eastern Pacific, with easterly surface winds and upper-level westerlies completing the circulation. Named for Sir Gilbert T. Walker (1868–1958), a British meteorologist who studied this circulation. Variations in this circulation are termed the **Southern Oscillation**.

warm air advection (or *warm advection*)—The transport of warm air from a region of high temperatures to a region of low (cold) temperatures by wind motion. Contrast **cold air advection**.

warm cloud—A cloud with temperatures exceeding 0°C throughout; as a result, the cloud contains only liquid cloud droplets, making it a **water cloud**.

warm front—Any nonoccluded **front**, or portion thereof, that moves in such a way that warmer air replaces colder air.

warm high—At a given level in the atmosphere, any **anticyclone** that is warmer at its center than at its periphery. Opposite of a **cold high**.

warm low—At a given level in the atmosphere, any **cyclone** that is warmer at its center than at its periphery. Opposite of a **cold low**. Examples include **hurricanes** and **heat lows**.

warm occlusion (or *warm occluded front*)—A type of **occluded front** resulting when the coldest air lies ahead of the **warm front**. This occlusion occurs when the original **cold front** is forced aloft at the warm frontal surface. At the earth's surface, the coldest air is replaced by less-cold air. Most often found in the Pacific Northwest. Compare with a **cold occlusion**.

warm pool—Any large-scale mass of warm air that appears as a ''pool'' of relatively warm air surrounded by colder air. Opposite of a **cold pool**.

warm sector—That area, within the circulation of a **wave cyclone**, where the warmest air typically is found. It lies between the **cold front** and **warm front** of the storm; and, in the typical case, the warm sector continually diminishes in size and ultimately disappears (at the surface) as the result of **occlusion**.

warm tongue—A pronounced poleward extension or protrusion of relatively warm air.

warm-core high—Same as a **warm high**.

warm-core low—Same as a **warm low**.

warning—A cautionary statement issued by the U.S. National Weather Service indicating that a specified hazardous weather or hydrologic event is imminent or actually occurring. Contrast with **outlook**, **watch**, and **weather advisory**. The intention of these warnings is to warn the public to take immediate appropriate action for personal safety. Types are **flash flood warning, hurricane warning**, **severe thunderstorm warning**, **tornado warning**, **winter storm warning**.

warning stage—The **stage**, or river level, on a fixed **river gauge**, at which **flood warnings** or **river forecasts** should be issued if adequate precautionary measures are to be taken before **flood stage** is reached.

watch—A cautionary statement issued by the U.S. National Weather Service indicating that atmospheric conditions are favorable for the development of a particular weather or hydrologic phenomenon (usually hazardous). Contrast with **outlook**, **weather advisory**, and **warning**. The intention is that these watches are to be issued with sufficient lead time to alert the public to make plans for personal safety if warranted. Watches are issued for **flash floods, hurricanes**, **severe thunderstorms**, **tornadoes**, **winter storms**.

water—Dihydrogen oxide (molecular formula H_2O); the word is used ambiguously to refer to the chemical compound in general and to its liquid phase.

water balance—Inventory of water based on the principle that, during a certain time interval, the total water gain to a given **catchment area** or body of water must equal the total water loss plus the net change in storage in the catchment.

water budget—A method for accounting for the amount (or volume) of water flowing into or out of a **reservoir**, as well as any changes in the stored amount of water in that reservoir.

water cloud—Any cloud composed entirely of liquid water drops; hence a **warm cloud**; to be distinguished from an **ice crystal cloud** and from a **mixed cloud**.

water content—The liquid water present within a sample of snow (or soil), usually expressed in percent by weight.

water cycle (or ***hydrologic cycle***)—The recurring succession of stages through which the water substance (H_2O) passes within the planet earth system at and near the earth's surface: **vaporization** from the land, sea, or inland waters; **condensation** (and **deposition**) to form clouds, **precipitation**, **accumulation** in the soil, sediment, or bedrock or in wetlands, lakes, or rivers, and **vaporization** (and **sublimation**) into the atmosphere.

water equivalent—The depth of water that would result from the melting of the **snowpack** or of a snow sample. Thus, the water equivalent of a new **snowfall** is the same as the amount of precipitation represented by that snowfall.

water mass—A vaguely homogeneous body of water having its source in a particular region of the ocean. It is usually identified by its **T–S curve** or chemical content, and usually consists of a mixture of several **water types**. The corresponding meteorological concept is **air mass**.

water table—The surface separating the upper layer of nonsaturated soil or sediment (**zone of aeration**) and the lower layer of saturated soil or sediment (**zone of saturation**), where the pressure is equal to the atmospheric pressure.

water type—**Seawater** of a specified temperature and **salinity**, and, hence, defined by a single point on a **temperature–salinity diagram**.

water vapor—Water substance (H_2O) in the gaseous phase; also known as *aqueous vapor*, or simply **vapor**. This very effective **greenhouse gas** is found primarily in the lower troposphere, with concentrations rarely exceeding 4% by volume.

water vapor image—An image from a satellite that senses **infrared radiation** at those wavelengths (typically near 6.7 micrometers) emitted by **clouds** and **water vapor** in the earth's atmosphere; because water also absorbs at these wavelengths, the images can show only flow patterns in the midtroposphere near the top of the bulk of the atmospheric water.

water year (or *hydrologic year*)—Any 12-month period, usually selected to begin and end during a relatively **dry season**, used as a basis for processing **streamflow** and other hydrologic data. The period from 1 October to 30 September is most widely used in the United States.

watercourse—The natural or artificial channel through or along which water may flow.

watershed—Same as **river basin**.

waterspout—Usually, a tornadolike rotating column of air (**whirlwind**) under a parent **cumuliform cloud** occurring over water; waterspouts are most common over tropical and subtropical waters and tend to dissipate upon reaching shore.

watt (W)—The derived unit for **power** in the SI units; named for James Watt (1736–1819), a Scottish engineer: 1 watt is equivalent to 1 joule per second.

wave—(1) Very generally, any pattern with some roughly identifiable periodicity in time and/or space. This applies, in meteorology, to atmospheric waves in the horizontal flow pattern (e.g., **Rossby wave**, **long wave**, **short wave**, **cyclone wave**). (2) Popularly used as a synonym for "surge" or "influx," as in **heat wave** or **cold wave**. (3) Undulations on the surface of a body of water; often of an oscillator nature.

wave cloud—An **orographic cloud**, at the crest of a **stationary wave**, formed in an airflow crossing a topographic barrier such as a range of hills or mountains.

wave crest—A point of maximum displacement of a **wave** above a mean value; the highest point of a wave. Opposite of a wave **trough**.

wave cyclone—A **cyclone** that forms and moves along a **front**; the circulation about the cyclone center tends to produce a wavelike deformation of the front.

wave depression—Same as **wave cyclone**.

wave disturbance—In **synoptic meteorology**, same as **wave cyclone**, but usually denoting an early stage in the development of a wave cyclone, or a poorly developed one.

wave group—A series of waves with a minimum variation in the direction, **wavelength**, and **wave height**.

wave height—The vertical distance between **wave crest** and **trough** in adjacent waves.

wave period—See **period** (1).

wave pole—A device for measuring sea surface waves. It consists of a weighted pole below which a disk is suspended at a depth sufficiently deep for the wave motion associated with **deep-water waves** to be negligible.

wave propagation—The process involving the transfer of a disturbance from one point to another in a medium without material **transport** of the medium. Displacement of waves through water or along the water surface.

wave recorder—An instrument for recording ocean waves. Most wave recorders are designed for recording **wind waves**, that is, waves of periods up to about 25 seconds, but some are designed to record waves of longer periods such as **tsunamis** or **tides**.

wave system—In ocean wave studies, a group of waves having the same height, direction, and length. Ocean surface waves are generally composed of a number of superimposed wave systems.

wave theory of cyclones—A theory of **cyclone** development based upon the principles of wave formation on an **interface** between two **fluids**. In the atmosphere, a cyclone would form along a **front**, taken as such an interface between different **air masses**. Compare with **convective theory of cyclogenesis**.

wave train—A series of waves traveling in the same direction.

wavelength—The distance between corresponding points (e.g., crest to crest) of two consecutive waves measured parallel to wave propagation. The wavelength is often used to describe wave phenomena characteristics, especially **electromagnetic radiation**; for example, **ultraviolet**, **visible**, and **infrared** radiation can be distinguished from one another by their characteristic wavelengths.

weather—(1) State of the atmosphere, mainly with respect to its effect upon life and human activities at a particular time, as defined by the various **weather elements**. As distinguished from **climate**, weather consists of the short-term (minutes to weeks) variations of the atmosphere. Popularly, weather is thought of in terms of temperature, humidity, precipitation, cloudiness, visibility, and wind. (2) In **aviation** and **synoptic weather observations**, weather constitutes those past (since the past observation) and present **atmospheric phenomena** of a significant nature, to include **tornadoes**, **thunderstorms**, **precipitation**, and **obstructions to vision**.

weather advisory—Meteorological information issued to alert the public when actual or expected weather conditions are of a nature that may cause general inconvenience or concern but do not constitute a hazard. See **advisory forecast**. Contrast with **outlook**, which may have a longer lead time, and the **watch** and **warning** messages, which alert or warn the public to weather events that could constitute life-threatening hazards.

weather analysis—The diagnostic study of the **synoptic** observational data using charts in order to represent the state of the atmosphere by **fronts**, and **isopleths** of various field properties (e.g., **isobars**, **isotherms**, and **contours**).

weather element—Any one of the observable properties or conditions of the atmosphere specifying the physical state of **weather** at a given place for any particular moment (e.g., temperature, humidity, precipitation).

weather forecast—A statement of expected meteorological conditions for a specific period and for a specific area; a **prognosis**.

weather map—Popularly, any graphical means of showing the distribution of meteorological conditions over a given (usually extensive) area of the earth's surface. See **constant height** and **constant pressure charts**.

weather modification—In general, any effort to alter artificially the natural phenomena of the atmosphere—for example, constructing windbreaks, dissipating fog, preventing frost, and stimulating precipitation.

weather observation—In general, an evaluation of one or more **weather elements** (e.g., temperature, dewpoint, precipitation, visibility, cloudiness) that describe the state of the atmosphere either at the earth's surface or aloft. **Surface weather observations** and **upper-air observations** are the major categories with a number of subtypes, but separate from these are such categories as radar meteorological observations, satellite observations, and solar radiation observations.

weather radar—An adaptation of **radar** for meteorological purposes. The scattering of **electromagnetic** waves, at wavelengths of a few millimeters to several centimeters, by raindrops and cloud drops is used to determine the distance, size, shape, location, motion, and phase (liquid and solid), as well as the intensity of the precipitation. Another application is in the detection of clear-air phenomena through scattering by insects and birds. See **Doppler radar**.

weather station—In a restricted sense, a location where meteorological observations of one kind or another are taken; more generally, the term also includes service offices that prepare weather maps and charts, issue forecasts and warnings, prepare and disseminate climatological information, and issue weather briefings to pilots.

weather warning—See **warning**.

weather watch—See **watch**.

weathering—The mechanical, chemical, or biological action of the atmosphere, **hydrometeors**, and suspended impurities on the form, color, or constitution of exposed material such as bedrock, roads, buildings. Contrast with **erosion** where material **transport** operates.

weighing rain gauge—A **recording rain gauge** for measuring and recording rainfall amounts by collecting the falling rain and weighing the **catch** by a weighing balance and recording the reading on a graph. The scale of the graph is calibrated such that the weight of the catch is recorded in conventional depth units (inches or millimeters). This instrument can determine the **precipitation intensity** by analyzing the record trace.

West Australia Current—The seasonal ocean current of the South Indian Ocean flowing along the west coast of Australia. In the Southern Hemisphere summer it flows northward, then curves toward the west to join the **South Equatorial Cur-**

rent. During Southern Hemisphere winter the West Australia Current flows southward.

West Greenland Current—The warm ocean current flowing northward along the west coast of Greenland into Davis Strait. It is a continuation of the **East Greenland Current**. Part of the **West Greenland Current** turns around when approaching the Davis Strait and joins the **Labrador Current**; the rest rapidly loses its character as a warm current as it continues into Baffin Bay.

west wind drift—Same as the surface layer of the **Antarctic Circumpolar Current**.

westerlies—(1) Specifically: the dominant west-to-east motion of the atmosphere, centered over the middle latitudes of both hemispheres. At the earth's surface, the *westerly belt* (or *west-wind belt*, etc.) extends, on the average, from about 35° to 65° latitude. At upper levels, the westerlies extend farther equatorward and poleward. The equatorward boundary is fairly well defined by the **subtropical high-pressure belt**; the poleward boundary is quite diffuse and variable. (2) Generally, any winds with components from the west; contrast with **easterlies**.

westerly wave—An atmospheric wavelike disturbance embedded in the midlatitude **westerlies**. Contrast with a **tropical easterly wave**.

western boundary current—The generic term for a relatively strong and narrow flow of ocean water (current) that runs deep along the western edge of a major ocean basin. Examples include the **Gulf Stream System** and the **Kuroshio System**.

wet adiabat—Same as **saturation adiabat**.

wet bulb—Contraction of either **wet-bulb temperature** or **wet-bulb thermometer**.

wet climate—A climate whose vegetation is of the **rainforest** type.

wet snow—Deposited snow that contains a great quantity of liquid water, or has a large **water equivalent**.

wet-adiabatic lapse rate—Same as **saturation-adiabatic lapse rate**.

wet-bulb depression—The difference in degrees between the **dry-bulb temperature** and the **wet-bulb temperature**; used in most **psychrometric tables**.

wet-bulb temperature—The temperature read from the **wet-bulb thermometer**. The temperature an **air parcel** would have if cooled adiabatically to saturation at constant pressure by evaporation of water into it, all **latent heat** of vaporization being supplied by the parcel. Contrast with **dry-bulb temperature**.

wet-bulb thermometer—In a **psychrometer**, the thermometer that has the wet, muslin-covered bulb and therefore measures **wet-bulb temperature**.

whaleback cloud—An elongated **lenticular cloud**.

whirlpool—A small-scale circulation feature where water moves rapidly and vigorously in a circular path; similar to an **eddy** or water **vortex**.

whirlwind—General term for a small-scale, rotating column of air; more specific terms are **dust whirl**, **dust devil**, **waterspout**, and **tornado**.

whistler—An electromagnetic signal of radio frequency that is sometimes produced by **lightning discharges**.

white dew—An occurrence of **dew** that has frozen as the result of a fall in temperature to below freezing after the original formation of the dew; contrast with **hoarfrost**, where ice on a surface results from **deposition** of atmospheric moisture onto the surface.

white frost—A relatively heavy coating of **hoarfrost**. A white frost is less damaging to vegetation than a **black frost** (or killing frost) for at least two reasons: (*a*) it tends to insulate the plant from further cold; and (*b*) it releases **latent heat** of fusion (albeit slight) to the environment.

whitebody—A hypothetical body whose surface absorbs no **electromagnetic radiation** of any wavelength; an idealization exactly opposite to that of the **blackbody**.

whitecap—A white patch of foam on the breaking **crest** of a wave being generated by the wind.

whiteout—An atmospheric optical phenomenon of the polar regions in which the observer appears to be engulfed in a uniformly white glow as a result of a lack of contrast between an **overcast** sky and an unbroken **snow cover**. Neither shadows, horizon, nor clouds are discernible; sense of depth and orientation is lost; dark objects in the field of view appear to "float" at an indeterminable distance.

whole gale—(1) In storm-warning terminology, a wind of 48–63 knots (55–72 mph). (2) The former name in the **Beaufort wind scale** (*Beaufort force number 10*) that is presently called **storm**, a wind whose speed is from 48 to 55 knots (55 to 63 mph).

whole-gale warning—A **warning**, for marine interests, of impending winds of 48–63 knots (55–72 mph). The storm-warning signals for this condition are (*a*) one square red flag with black center by day and (*b*) two red lanterns by night.

Wien's displacement law—One of the **blackbody radiation laws** that states that the wavelength of maximum radiation intensity for a **blackbody** is inversely proportional to the absolute temperature of the radiating blackbody. This relationship was defined in 1894 by Wilhelm Wien (1864–1928), a German physicist.

willywaw—A **bora**-type wind in Alaska.

willy-willy—A severe **tropical cyclone** (with winds 33 knots or greater) in Australia, especially in the southwest.

wilting point—Moisture content of the soil at which leaves of plants growing in that soil would suffer from a collapse of cellular tissue and die by wilting because of moisture stress; the wilting point is expressed as a percentage of mass of dry soil.

wind—Air in motion relative to the surface of the earth. Since vertical components of atmospheric motion are relatively small especially near the surface of the earth, meteorologists tend to use the term to denote almost exclusively the horizontal component. Vertical winds are always identified as such.

wind advisory—Issued by the National Weather Service to make the public aware of anticipated sustained **wind speeds** of 30 miles per hour or greater lasting one hour or longer, or **gusts** to 45 miles per hour for any duration. See **high wind warning**.

wind arrow—Same as **wind shaft**.

windbreak—Any device or barrier designed to obstruct wind flow and intended for protection of the immediate downwind area against any ill effects of wind. Installations of this type include **shelterbelts**, **snow fences**, and **rain-gauge shields**.

windburn—(1) A superficial inflammation of the skin, analogous to **sunburn**, caused by exposure to wind, especially a hot dry wind, inducing a dilatation of the surface blood vessels. (2) Injury to plant foliage, caused by strong, hot, dry winds.

wind chill—That part of the total cooling of a body caused by air motion. See also **wind chill factor** and **wind-chill equivalent temperature**.

wind chill advisory—Issued by the National Weather Service to inform the public that **wind-chill equivalent temperatures** are generally expected to reach –20°F or colder for a sustained time interval. **Wind speeds** are expected to be at least 10 miles per hour. Mitigating circumstances such as strong sunshine and acclimation in cold climates, may require colder wind chill thresholds.

wind chill equivalent temperature (**WET**)—An air temperature index (in units of degrees Fahrenheit or Celsius) that attempts to approximate the **sensible heat** loss from exposed skin caused by the combined effect of low air temperature and wind; defined as the ambient air temperature that would produce, under essentially **calm** wind conditions (4 knots), the same heat loss (**wind chill factor**) as the actual combination of ambient air temperature and wind speed. Only when the wind is essentially calm, will the actual air temperature be the same as the wind chill equivalent temperature.

wind chill factor—The cooling effect of any combination of temperature and wind, expressed as the loss of body heat in kilogram-calories per hour per square meter of skin surface; the numeric values of this factor are used in public weather forecasts issued by Environment Canada. The term is commonly used for **wind chill equivalent temperature**.

wind chill warning—Issued by the National Weather Service to warn the public of the potentially life-threatening situation produced by **wind chill equivalent temperatures** expected to reach –50°F or colder and **wind speeds** of 10 miles per hour or greater.

wind direction—The direction from which the wind is blowing; often measured by a **wind vane**. The wind direction may be reported in terms of the points of a compass (e.g., a north wind from the north) or as a **bearing** (e.g., a south wind reported as 180°).

wind divide—A semipermanent feature of the atmospheric circulation (usually a high-pressure **ridge**) on opposite sides of which the prevailing **wind directions** differ greatly.

wind drift—Same as **wind-driven current**.

wind factor—In oceanography, the ratio of the velocity of an ocean **surface current** to the velocity of the wind.

wind pressure—The total **force** exerted upon a structure by wind on a flat surface; the wind pressure on the windward side is the **dynamic pressure**, a function of the square of the **wind speed**.

wind profile—Graphical representation of **wind speed** and/or **wind direction** as a function of altitude or distance.

wind profiler—A three beam **Doppler radar** aimed vertically in order to measure atmospheric winds above a station, to produce a **wind profile**, from which potentially dangerous **wind shear** could be detected.

wind rose—Any one of a class of diagrams designed to show the distribution of **wind direction** experienced at a given location over a considerable time interval; it thus shows the prevailing **wind direction**. Usually, these diagrams are star-shaped with 4–36 lines (depending upon the desired number of directional categories) emanating from a center, with the line lengths depicting the wind frequency from a given compass direction.

wind scoop—A saucerlike depression in the **snow cover** near obstructions such as trees, houses, and rocks, caused by the eddying action of the deflected wind.

wind shaft (or *wind arrow*)—A short straight line terminating at the **station model** on a synoptic chart and representing the direction from which the wind blows.

wind shear—The local variation of the **wind vector** or any of its components in a given direction; hence, wind shear can have *directional shear*, *speed shear*, or a combination of both.

wind shield—See **rain-gauge shield**.

wind shift—A sudden change of **wind direction**. According to United States observation procedure, this condition would be reported if a change in wind direction of 45° or more took place in less than 15 minutes.

wind sock—A tapered fabric sleeve, shaped like a truncated cone and pivoted at its larger end on a standard, for the purpose of indicating **wind direction**. Since the air enters the fixed end, the small end of the cone points away from the wind.

wind speed—Ratio of the distance covered by the air to the time taken to cover it; typically measured with an **anemometer**. The "instantaneous speed" or, more briefly, the "speed," corresponds to the case of an infinitely small time interval. The "mean speed" corresponds to the case of a finite time interval. See **gust** and **peak wind speed**.

wind stress—The drag or tangential force per unit area exerted upon the earth's surface by moving air in the **surface boundary layer**.

wind vane—An instrument used to indicate **wind direction**.

wind vector—A **vector** quantity, usually portrayed by an arrow displayed on a chart to represent the **wind velocity**; the arrow is oriented to point in the same direction as the **wind direction**, while its length indicates the **wind speed**.

wind velocity—A **vector** quantity describing the characteristics of wind motion in terms of **wind speed** and **wind direction**; a **wind vector** is a graphical representation of wind velocity.

wind wave—A wave resulting from the action of wind on a water surface. While the wind is acting on it, it is a **sea** (2), thereafter, a **swell**.

winds aloft—Generally, the **wind speeds** and **directions** at various levels in the atmosphere above the domain of **surface weather observations**; made by **pilot balloons**, **radiosondes**, **AIREPS**, or cloud tracking by satellites.

wind-driven current (also called *drift current* or *wind drift*)—Motion of a body of water due to the action of the wind on its surface. Contrast with **density** and **geostrophic currents**.

wind-driven oceanic circulation—Horizontal water motion of the ocean due to the **stress** exerted on its surface by the wind.

wind-shift line—A line or narrow zone along which an abrupt change of **wind direction** occurs.

windward—The direction or the side of any object (e.g., island, mountain, or region) that faces the prevailing wind. Opposite of **leeward**.

winter—(1) The coldest **season** of the year; the "low sun" season during which the sun is over the opposite hemisphere; the "hibernal" season. Popularly and for most meteorological purposes, winter is taken to include December, January, and February in the Northern Hemisphere; and, in the Southern Hemisphere, June, July, and August; the reverse of **summer**. (2) Astronomically in the Northern Hemisphere, the period extending from the **winter solstice**, about 22 December, to the **vernal equinox**, about 21 March.

winter monsoon—The seasonal wind regime (**monsoon**) of continental origin that blows toward the ocean in the low sun (winter) season. Opposite of the **summer monsoon**.

winter solstice—The point on the **celestial sphere** where the sun reaches it greatest distance (or **declination**) north or south from the **celestial equator** for the Southern or Northern Hemisphere, respectively. Opposite of the **summer solstice**. In the Northern Hemisphere, the time of this occurrence is approximately 22 December, when the sun appears overhead on the **Tropic of Capricorn**.

winter storm—A term often used in an operational weather discussion, such as a forecast, to describe a weather event with hazardous conditions associated with significant quantities of **freezing** or **frozen precipitation**, or the combined effects associated with winter precipitation, strong winds, and cold temperatures. See **winter storm watch** and **winter storm warning**.

winter storm warning—Issued by the National Weather Service to warn the public that **heavy snow** and some winds are imminent or very likely, perhaps in combination with **ice pellets** (**sleet**) and/or **freezing rain** or **freezing drizzle**. Winter storm warnings are usually issued for up to a 12-hour duration, but can be extended to 24 hours. A warning is used for conditions posing a threat to life or property.

winter storm watch—Issued by the National Weather Service to alert the public that conditions are favorable for the development of hazardous **weather elements** such as **heavy snow** and/or **blizzard** conditions, or significant accumulations of **freezing rain** or **ice pellets** (**sleet**). Watches are usually issued 12–48 hours in advance of the expected event.

winter weather advisory—Issued by the National Weather Service to advise the public of the expectation of a mixture of precipitation, such as snow, **ice pellets** (**sleet**), **and freezing rain** (or **freezing drizzle**). Advisories are conditions less serious than warnings that cause significant inconvenience and, if caution is not exercised, can be a threat to life and property.

work—An **energy** form arising from the displacement of an object by a **force**; mathematically, the product of a displacement and the component of the applied force in the direction of that displacement. The unit of work is the same as **energy**, or specifically, **joule**.

World Meteorological Organization (**WMO**)—Specialized agency of the United Nations Organization for coordinating, standardizing, and improving meteorological activities throughout the world and for encouraging the efficient exchange of information between countries, in the interest of various human activities.

World Weather Watch (**WWW**)—A coordinated international system for the collection, analysis, and distribution of weather information under the auspices of the **World Meteorological Organization**. This program includes the Global Observing System, the Global Data-Processing System, and the Global Telecommunication System.

X

XBT—Abbreviation for expendable **bathythermograph**.

X-rays—Very energetic **electromagnetic radiation** with wavelengths intermediate between 0.01 and 10 nanometers (0.1–100 **Ångstroms**) or between **gamma rays** and **ultraviolet radiation**. Essentially all X-rays from space are absorbed in the earth's upper atmosphere.

Y

year—(1) The period during which the earth completes one revolution around the sun. This has several interpretations. (*a*) *Sidereal year*: time of true revolution around the sun—that is, the time it takes the earth (as seen from the sun) to reappear at the same fixed star, equal to 365.2564 mean solar days or 365 days, 6 hours, 9 minutes, 10 seconds. (*b*) *Tropical year* (also called *mean solar year*, *ordinary year*): the time measured from one vernal equinox to the next—that is, the apparent revolution of the sun through the zodiac, equal to 365.2422 mean solar days or 365 days, 5 hours, 48 minutes, 46 seconds. (This is not constant but decreases by about 5 seconds in one thousand years.) (*c*) *Calendar year*; fixed by the Gregorian calendar of 365 days in ordinary years and 366 in leap years. (2) Any arbitrary 12-month period selected for a special purpose, such as the **water year**.

yellow snow—Snow given a golden or yellow appearance by the presence in it of pine, cypress pollen, or anthropogenic material or animal-produced material. Compare with **brown snow**.

Z

zenith—That point, on any given observer's **celestial sphere**, that lies directly overhead; the point that is elevated 90° from all points on a given observer's astronomical **horizon**. Diametrically opposite the zenith is the observer's **nadir**.

zenith angle—The angular distance on the **celestial sphere**, between the observer's **zenith** and the celestial body, measured along the **great circle** passing through the zenith and the body in question. The zenith angle is 90° minus the **altitude angle**.

zenithal rains—In the Tropics or subtropics, **rainy seasons** that recur annually or semiannually at about the time that the sun is most nearly overhead (at **zenith**). Within a few degrees of the equator, two such periods of heavy rains may occur annually, the **equinoctial rains**. Farther from the equator, these may blend into a single annual summer rainy season.

zephyr—Any soft, gentle breeze.

Zephyros—The ancient Greek name for the west wind, which generally is light and beneficial in the eastern Mediterranean. On the **Tower of the Winds**, it is represented by a youth wearing only a light mantle, with a skirt filled with flowers.

zodiac—The path of the sun during the course of a year as it appears to move through successive star groups or constellations; that is, the band of the **celestial sphere**, 16° in width, through which the **ecliptic** runs centrally. At all times this band of the heavens contains the sun, the moon, and the principal planets.

zodiacal light—A faint cone of light extending upward from the horizon aligned along the **ecliptic** (or **zodiac**). Its source appears to be sunlight scattered by interplanetary debris found close to the plane of the ecliptic. It is seen best from most midlatitude locales after sunset near the **vernal equinox** or before sunrise near the **autumnal equinox**.

zonal—In meteorology, latitudinal along a parallel of **latitude** (or zone); easterly or westerly; opposed to **meridional**.

zonal flow—In meteorology, the flow of air along a **latitude** circle; more specifically, the latitudinal (east or west) component of existing flow.

zonal wind—The wind, or wind component, along the local parallel of latitude, as distinguished from the **meridional wind**. In a horizontal coordinate system fixed locally with the *x* axis directed eastward and the *y* axis directed northward, the zonal wind component is positive if it blows from the west and negative if from the east.

zone of aeration—The region of soil, rock, or sediment above a **water table** where the pore spaces contain air as well as water. Contrast with **zone of saturation**.

zone of saturation—The soil, rock, or sediment beneath the **water table**. All pore spaces in the zone of saturation are filled with water, in contrast to the pore spaces in the **zone of aeration** above the water table, which may contain considerable air.

"Zulu" (Z) time—For practical purposes, the same as **Universal Coordinated Time (UTC)**. The notation formerly used to identify time *Greenwich Mean Time*. The word "Zulu" is notation in the phonetic alphabet corresponding to letter "Z" assigned to the time zone on the Greenwich Prime Meridian.

Appendix: The SI or International System of Units

The International System of Units (abbreviated SI, for Système Internationale d' Unités), the internationally approved version of the metric measurement system, is based upon seven dimensionally independent ***basic or fundamental units***. Other quantities, defined in terms of these fundamental units, may have approved names.

Sources: *Bull. Amer. Meteor. Soc., 55*, 926–930. August, 1974.
U.S. Dept. of Commerce, 1972: *The International System of Units (SI)*. Natl. Bur. Stand. (U.S.) Spec. Publ. 330. Washington, DC. 42 pp.

SI Base (or Fundamental) Units

Physical Quantity	Name of Unit	Symbol
Length	meter	m
Mass	kilogram	kg
Time	second	s
Electric current	ampere	A
Thermodynamic temperature	kelvin	K
Amount of substance	mole	mol
Luminous intensity	candela	cd

Examples of SI Derived Units expressed in terms of Base Units

Physical Quantity	Name of Unit	Definition of Unit
Area	square meter	m^2
Volume	cubic meter	m^3
Speed, velocity	meter per second	$m\ s^{-1}$

continued

Examples of SI Derived Units expressed in terms of Base Units (*Continued*)

Physical Quantity	Name of Unit	Definition of Unit
Acceleration	meter per second squared	$m\ s^{-2}$
Density, mass density	kilogram per cubic meter	$kg\ m^{-3}$
Specific volume	cubic meter per kilogram	$m^3\ kg^{-1}$
Concentration (amount)	mole per cubic meter	$mol\ m^{-3}$
Divergence	per second	s^{-1}
Vorticity	per second	s^{-1}
Activity (radioactive)	1 per meter	s^{-1}
Wavenumber	1 per meter	m^{-1}
Geopotential, dynamic height	meter squared per second squared	$m^2\ s^{-2}$
Luminance	candela per square meter	$cd\ m^{-2}$

SI Derived Units with special names

Physical Quantity	Name of Unit	Symbol	Definition of Unit
Frequency	hertz	Hz	s^{-1} (cycle per second)
Force	newton	N	$m\ kg\ s^{-2}$
Pressure	pascal	Pa	$kg\ m^{-1}\ s^{-2} = N\ m^{-2}$
Energy (work, heat quantity)	joule	J	$kg\ m^2\ s^{-2} = N\ m$
Power, radiant flux	watt	W	$kg\ m^2\ s^{-3} = J\ s^{-1}$
Electric charge	coulomb	C	$A\ s$
Electric potential	volt	V	$kg\ m^2\ s^{-3}\ A^{-1} = J\ A^{-1}\ s^{-1} = W\ A^{-1}$
Electric resistance	ohm	Ω	$kg\ m^2\ s^{-3}\ A^{-2} = V\ A^{-1}$

Examples of SI Derived Units expressed by means of special names

Physical Quantity	Name of Unit	Definition of Unit
Dynamic viscosity	pascal second	$Pa\ s = kg\ m^{-1}\ s^{-1}$
Moment of force	meter newton	$N\ m = m^2\ kg\ s^{-2}$
Surface tension	newton per meter	$N\ m^{-1} = kg\ s^{-2}$
Heat flux density	watt per square meter	$W\ m^{-2} = kg\ s^{-3}$
Entropy	joule per kelvin	$J\ K^{-1} = m^2\ kg\ s^{-2}\ K^{-1}$
Gas constant, universal	joule per kilogram per kelvin [*or*, joule per kilomole per kelvin]	$J\ kg^{-1}\ K^{-1} = m^2\ s^{-2}\ K^{-1}$ $J\ kmol^{-1}\ K^{-1}$
Specific heat capacity	joule per kilogram per kelvin	$J\ kg^{-1}\ K^{-1} = m^2\ s^{-2}\ K^{-1}$
Specific energy	joule per kilogram	$J\ kg^{-1} = m^2\ s^{-2}$
Thermal conductivity	watt per meter per kelvin	$W\ m^{-1}\ K^{-1} = m\ kg\ s^{-2}\ K^{-1}$

(continued)

Examples of SI Derived Units expressed by means of special names *(Continued)*

Physical Quantity	Name of Unit	Definition of Unit
Energy density	joule per cubic meter	$J\ m^{-3} = kg\ m^{-1}\ s^{-2}$
Electric field strength	volt per meter	$V\ m^{-1} = m^{-1}\ kg\ s^{-2}\ s^{-1}$

SI Supplementary Units

Physical Quantity	Name of Unit	Symbol
Plane angle	radian	rad
Solid angle	steradian	sr

Examples of SI Derived Units formed by using Supplementary Units

Physical Quantity	Name of Unit	Symbol
Angular velocity	radian per second	$rad\ s^{-1}$
Angular acceleration	radian per second squared	$rad\ s^{-2}$
Radiant intensity	watt per steradian	$W\ sr^{-1}$
Radiance	watt per square meter per steradian	$W\ m^{-2}\ sr^{-1}$

In addition to the base unit of measurement for each dimensional quantity, the SI system employs a standard system of prefixes for larger or smaller units of measure, based upon multiples of ten. The following table lists the preferred prefixes used. For example, 1000 (= 10^3) meter is 1 kilometer (km), and 10^{-6} meter is 1 micrometer (mm).

Accepted SI prefixes and symbols for multiples and submultiples of units

Multiple	Prefix	Symbol	Submultiple	Prefix	Symbol
10^{12}	tera	T	10^{-1}	deci	d
10^{9}	giga	G	10^{-2}	centi	c
10^{6}	mega	M	10^{-3}	milli	m
10^{3}	kilo	k	10^{-6}	micro	μ
10^{2}	hecto	h	10^{-9}	nano	n
10^{1}	deka	da	10^{-12}	pico	p
			10^{-15}	femto	g
			10^{-18}	atto	a

References

Allsopp, J., R.J. Vavrek, and R.L. Holle, 1995: Is it going to rain today? Understanding the weather forecast. *The Earth Scientist, 12 (4),* 12–18.

American Geological Institute, 1962: *Dictionary of Geological Terms.* Anchor Books, Garden City, NY. 545 pp.

American Society of Heating, Refrigerating and Air-Conditioning Engineers, 1992: *Thermal Environmental Conditions for Human Occupancy.* ASHRAE Standards. ANSI/ASHRAE 55-1992. 20 pp.

Baker, B.B., Jr., W.R. Deebel, and R.D. Geisenderfer, 1966: *Glossary of Oceanography Terms.* 2d ed. U.S. Naval Oceanographic Office. Special Publication SP-35. U.S. Govt. Printing Office, Washington, DC. 204 pp.

Committee on Extension to the Standard Atmosphere, 1976: *U.S. Standard Atmosphere, 1976:* U.S. Government Printing Office, Washington, DC. 227 pp.

Department of the Air Force, 1994: *Surface Aviation Observations: Meteorological Aviation Report.* Air Force Manual 15-111. Vol. 2. U.S. Govt. Printing Office, Washington, DC.

Federal Aviation Administration, 1985: *Aviation Weather Services.* (AC 00-45C) U.S. Dept. of Transportation. U.S. Govt. Printing Office, Washington, DC.

Federal Aviation Administration, 1985: *Handbook of Applied Meterology.* D.D. Houghton, Ed., John Wiley & Sons, New York, NY. 1461 pp.

Huschke, R.E., ed., 1959: *Glossary of Meteorology.* American Meteorological Society, Boston, MA. 638 pp.

Lewis, R.P.W., 1991: *Meteorological Glossary.* 6th ed. Her Majesty's Stationary Office. 335 pp. [Available through Unipub, Lanham, MD.]

Morris, C., ed., 1992: *Academic Press Dictionary of Science and Technology.* Academic Press, San Diego, CA. 2432 pp.

National Aeronautics and Space Administration, 1988: *NASA Thesaurus.* Vol. 3., *Definitions.* NASA SP-7064 Science and Technical Information Division. National Aeronautics and Space Administration, Washington, DC. 142 pp.

National Aeronautics and Space Administration, 1994: *Looking at Earth from Space. Glossary of Terms.* (EP-302) Office of Mission to Planet Earth. National Aeronautics and Space Administration, Washington, DC. 76 pp.

National Oceanic and Atmospheric Administration, 1988: *Federal Meteorological Handbook No. 1: Surface Observations.* (FCM-H1-1988) U.S. Dept. of Commerce. U.S. Govt. Printing Office, Washington, DC.

National Weather Service, 1989: *Cooperative Station Observations. National Weather Service Observing Handbook No. 2.* National Oceanic and Atmospheric Administration. U.S. Dept. of Commerce. U.S. Govt. Printing Office, Washington, DC.

References

National Weather Service, 1992: *A Mariner's Guide To Marine Weather Services.* National Oceanic and Atmospheric Administration. U.S. Dept. of Commerce. U.S. Govt. Printing Office. Washington, DC. Single sheet folded pamphlet.

National Weather Service, 1993: *A Pilot's Guide to Aviation Weather Services.* National Oceanic and Atmospheric Administration. NOAA PA/93056. U.S. Dept. of Commerce. U.S. Supt. of Documents. Washington, DC. 17 pp.

National Weather Service, 1993: *"Hurricane!" A Familiarization Guide.* Revised edition. National Oceanic and Atmospheric Administration. NOAA PA/91001. U.S. Dept. of Commerce. U.S. Supt. of Documents, Washington, DC. 36 pp.

Nayler, J.L., 1964: *A Dictionary of Astronautics.* Hart Publishing Co., New York, NY. 316 pp.

Sandak, C.R., 1990: *A Reference Guide to Clean Air. Science, Technology and Society Reference Series.* Enslow Publishers, Inc., Hillside, NJ. 128 pp.

Thompson, S., 1994: *Wisconsin Media Guide to Frequently Used NWS Weather Terms.* Unpublished, National Weather Service Office, La Crosse, WI. 9 pp.

U.S. Dept. of Commerce, 1982: *National Weather Service Observers Handbook No. 1. Marine Surface Weather Observations.* National Oceanic and Atmospheric Administration. Supt. of Documents, Washington, DC.

U.S. Geological Survey, 1994: *Water Resources Data—Wisconsin.*

U.S. Naval Observatory, 1995: *The Astronomical Almanac.* U.S. Govt. Printing Office, Washington, DC.

Walker, P.M.B., ed., 1990: *Cambridge Air And Space Dictionary.* Cambridge University Press, New York, NY. 216 pp.

World Meteorological Organization, 1956: *International Cloud Atlas.* Vol. 1. World Meteorological Organization. Geneva, Switzerland. 155 pp.

World Meteorological Organization, 1987: *International Cloud Atlas.* Vol. 2. Rev. ed. World Meteorological Organization. Geneva, Switzerland. 212 pp.

World Meteorological Organization, 1992: *International Glossary of Hydrology.* 2d ed. Joint publication World Meteorological Organization and United Nations Education, Scientific and Cultural Organization, Geneva, Switzerland. 413 pp.

World Meteorological Organization, 1992: *International Meteorological Vocabulary.* 2d ed. WMO/OMM/BMO No. 182. World Meteorological Organization, Geneva, Switzerland. 784 pp.

Wurtele, M.G., 1985: Glossary and Units. *Handbook of Applied Meteorology.* D.D. Houghton, Ed., John Wiley & Sons, New York, NY. 1357–1368.